The Game of Pig

Probability and Expected Value

Teacher's Guide

This material is based upon work supported by the National Science Foundation under award numbers ESI-9255262, ESI-0137805, and ESI-0627821. Any opinions, findings, and conclusions or recommendations expressed in this publication are those of the authors and do not necessarily reflect the views of the National Science Foundation.

© 2009 by Interactive Mathematics Program. Some rights reserved. Permission is granted to reproduce and distribute all or any portion of the *IMP Teacher's Guide* for non-commercial purposes. Permission is also granted to make Derivative products, modifying the *IMP Teacher's Guide* content, for non-commercial purposes. Any Derivatives must carry the statement "This material is adapted from the IMP Teacher's Guide, © 2009 Interactive Mathematics Program. Some rights reserved."

®Interactive Mathematics Program is a registered trademark of Key Curriculum Press. ™IMP and the IMP logo are trademarks of Key Curriculum Press.

Key Curriculum Press
1150 65th Street
Emeryville, California 94608
email: editorial@keypress.com
www.keypress.com

First Edition Authors
Dan Fendel, Diane Resek, Lynne Alper, and Sherry Fraser

Contributors to the Second Edition
Sherry Fraser, Jean Klanica, Brian Lawler, Eric Robinson, Lew Romagnano, Rick Marks, Dan Brutlag, Alan Olds, Mike Bryant, Jeri P. Philbrick, Lori Green, Matt Bremer, Margaret DeArmond

Project Editors
Joan Lewis, Sharon Taylor

Consulting Editor
Mali Apple

Editorial Assistant
Juliana Tringali

Professional Reviewer
Rick Marks, Sonoma State University

Calculator Materials Editor
Christian Aviles-Scott

Math Checker
Christian Kearney

Production Director
Christine Osborne

Executive Editor
Josephine Noah

Textbook Product Manager
Tim Pope

Publisher
Steven Rasmussen

Contents

Introduction

The Game of Pig Unit Overview	v
Pacing Guides	vii
Materials and Supplies	xiii
Assessing Progress	xiv
Supplemental Activities Overview	xv

Activity Notes

Chance and Strategy	**1**
The Game of Pig	3
POW 3: A Sticky Gum Problem	5
Pig at Home	7
Pig Strategies	8
Waiting for a Double	10
The Gambler's Fallacy	13
Expecting the Unexpected	16
Coincidence or Causation?	19
What Are the Chances?	21
Paula's Pizza	26
0 to 1, or Never to Always	28
Pictures of Probability	**29**
Rug Games	30
Portraits of Probabilities	33
POW 4: Linear Nim	36
Mystery Rugs	38
The Counters Game	39
Rollin', Rollin', Rollin'	41
The Theory of Two-Dice Sums	44
Money, Money, Money	48
Two-Dice Sums and Products	50
In the Long Run	**52**
Spinner Give and Take	54
Pointed Rugs	58
POW 5: What's on Back?	60
Mia's Cards	65
A Fair Rug Game?	68

One-and-One	70
A Sixty Percent Solution	73
The Theory of One-and-One	75
Streak-Shooting Shelly	80
Spins and Draws	82
Aunt Zena at the Fair	85
Simulating Zena	86
The Lottery and Insurance—Why Play?	88
Martian Basketball	90
The Carrier's Payment Plan Quandary	95
A Fair Deal for the Carrier?	100
Simulating the Carrier	101
Another Carrier Dilemma	103
Analyzing a Game of Chance	**105**
The Game of Little Pig	106
Pig Tails	108
Little Pig Strategies	110
Continued Little Pig Investigation	114
Should I Go On?	117
The Best Little Pig	119
Big Pig Meets Little Pig	121
The Pig and I	123
Beginning Portfolio Selection	124
The Game of Pig Portfolio	125

Blackline Masters

Rug Games Blackline Master
1-Centimeter Graph Paper Blackline Master
1-Inch Graph Paper Blackline Master
1/4-Inch Graph Paper Blackline Master
In-Class Assessment
Take Home Assessment

Calculator Guide and Calculator Notes

Introduction

The Game of Pig Unit Overview

Intent

The second unit of Year 1 is built around a central problem: developing optimal strategies for winning a dice game called Pig. The mathematics in this unit centers on key ideas of probability and strategic thinking. Students will continue to develop their abilities to tackle substantial problems, to reason mathematically, and to communicate their thinking—that is, their understanding of what it means to do mathematics.

Mathematics

As an introduction to the probability concepts and skills needed to analyze the game of Pig, students work on a variety of problems involving chance occurrences. Through these experiences, they develop an understanding of the concept of expected value and learn to calculate expected value using an area model. They also encounter some real-life "games," such as buying insurance and playing the lottery, and discover that in such situations, expected value may not be the sole criterion for making a decision.

In the unit activities, students explore these important mathematical ideas:

- Learning what constitutes a "complete strategy" for a game and developing and analyzing strategies
- Calculating probabilities as fractions, decimals, and percents by emphasizing equally likely outcomes and by constructing mathematical models, including area models and tree diagrams
- Determining whether events are independent
- Using the idea of "in the long run" to develop the concept of expected value and calculating and interpreting expected values
- Solving problems involving conditional probability
- Making and interpreting frequency bar graphs
- Using simulations to estimate probabilities and compare strategies
- Comparing the theoretical analysis of a situation with experimental results
- Examining how the number of trials in a simulation affects the results

Progression

In *Chance and Strategy,* the unit opens with an introduction to the game of Pig. Students play the game, begin to get a feel for the variety of possible outcomes, and speculate informally about strategies for playing the game. In *Pictures of Probability,* they leave the game behind to focus on developing a quantitative understanding of, and area models for, probabilistic events. In *In the Long Run,* students conduct an in-depth exploration of the concept of expected value. Finally, in *Analyzing a Game of Chance,* they use the tools they have developed to analyze a simplified version of Pig and then return to an analysis of the original game.

Chance and Strategy

Pictures of Probability

In the Long Run

Analyzing a Game of Chance

Supplemental Activities

Assessing Progress

Pacing Guides

50-Minute Pacing Guide (30 days)

Day	Activity	In-Class Time Estimate
Chance and Strategy		
1	*The Game of Pig*	35
	POW 3: A Sticky Gum Problem	10
	Homework: *Pig at Home*	5
2	Discussion: *Pig at Home*	15
	Pig Strategies	30
	Homework: *Waiting for a Double*	5
3	Discussion: *Waiting for a Double*	25
	The Gambler's Fallacy	20
	Homework: *Expecting the Unexpected*	5
4	*The Gambler's Fallacy* (continued)	30
	Discussion: *Expecting the Unexpected*	15
	Homework: *Coincidence or Causation?*	5
5	Discussion: *Coincidence or Causation?*	15
	What Are the Chances?	35
	Homework: *Paula's Pizza*	0
6	Discussion: *Paula's Pizza*	15
	What Are the Chances? (continued)	30
	Homework: *0 to 1, or Never to Always*	5
7	Discussion: *0 to 1, or Never to Always*	10
Pictures of Probability		
	Rug Games	35
	Homework: *Portraits of Probabilities*	5
8	Discussion: *Portraits of Probabilities*	15
	Presentations: *POW 3: A Sticky Gum Problem*	20
	POW: Linear Nim	15

		Homework: *Mystery Rugs*	0
9		Discussion: *Mystery Rugs*	10
		The Counters Game	35
		Homework: *Rollin', Rollin', Rollin'*	5
10		Discussion: *Rollin', Rollin', Rollin'*	25
		The Theory of Two-Dice Sums	25
		Homework: *Money, Money, Money*	0
11		Discussion: *Money, Money, Money*	20
		The Theory of Two-Dice Sums (continued)	30
		Homework: *Two-Dice Sums and Products*	0
12		Discussion: *Two-Dice Sums and Products*	15
In the Long Run			
		Spinner Give and Take	35
		Homework: *Pointed Rugs*	0
13		Discussion: *Pointed Rugs*	15
		Presentations: *POW 4: Linear Nim*	20
		POW 5: What's on Back?	10
		Homework: *Mia's Cards*	5
14		Discussion: *Mia's Cards*	30
		Reflection: Expected Value and the Unit Problem	5
		POW 5: What's on Back?	10
		Homework: *A Fair Rug Game?*	5
15		Discussion: *A Fair Rug Game?*	20
		One-and-One	25
		Homework: *A Sixty Percent Solution*	5
16		Discussion: *A Sixty Percent Solution*	10
		The Theory of One-and-One	40
		Homework: *Streak-Shooting Shelly*	0
17		Discussion: *Streak-Shooting Shelly*	15
		Spins and Draws	35
		Homework: *Aunt Zena at the Fair*	0
18		Presentations: *POW 5: What's on Back?*	20

	Discussion: *Aunt Zena at the Fair*	15
	Simulating Zena	15
	Homework: *The Lottery and Insurance—Why Play?*	0
19	Discussion: *The Lottery and Insurance—Why Play?*	10
	Martian Basketball	30
	Homework: *The Carrier's Payment-Plan Quandary*	10
20	*Martian Basketball* (continued)	15
	Discussion: *The Carrier's Payment-Plan Quandary*	30
	Homework: *A Fair Deal for the Carrier?*	5
21	Discussion: *A Fair Deal for the Carrier?*	25
	Simulating the Carrier	25
	Homework: *Another Carrier Dilemma*	0
22	Discussion: *Another Carrier Dilemma*	10
Analyzing a Game of Chance		
	The Game of Little Pig	35
	Homework: *Pig Tails*	5
23	Discussion: *Pig Tails*	15
	Little Pig Strategies	30
	Homework: *Continued Little Pig Investigation*	5
24	Discussion and Activity: *Continued Little Pig Investigation*	45
	Homework: *Continued Little Pig Investigation* (continued)	5
25	Discussion: *Continued Little Pig Investigation*	45
	Homework: *Should I Go On?*	5
26	Discussion: *Should I Go On?*	15
	The Best Little Pig	25
	Homework: *Big Pig Meets Little Pig*	10
27	Discussion: *Big Pig Meets Little Pig*	40
	Homework: *The Pig and I*	10
28	*Beginning Portfolio Selection*	20
	The Game of Pig Portfolio	30
29	*In-Class Assessment*	50

	Homework: *Take-Home Assessment*	0
30	Discussion: *In-Class Assessment* and *Take-Home Assessment*	30
	Unit Reflection	20

90-minute Pacing Guide (17 days)

Day	Activity	In-Class Time Estimate
Chance and Strategy		
1	*The Game of Pig*	35
	Waiting for a Double	35
	POW: A Sticky Gum Problem	10
	Homework: *Expecting the Unexpected*	5
	Homework: *Pig at Home*	5
2	Discussion: *Pig at Home*	5
	Pig Strategies	25
	Waiting for a Double (continued)	20
	The Gambler's Fallacy	40
	Homework: *Coincidence or Causation?*	0
3	Discussion: *Expecting the Unexpected*	15
	Discussion: *Coincidence or Causation?*	15
	What Are the Chances?	60
	Homework: *Paula's Pizza*	0
	Homework: *0 to 1, or Never to Always*	0
Pictures of Probability		
4	Discussion: *Paula's Pizza*	10
	Discussion: *0 to 1, or Never to Always*	15
	Rug Games	35
	Portraits of Probabilities	30
	Homework: *Mystery Rugs*	0
5	Discussion: *Mystery Rugs*	15
	Presentations: *POW: A Sticky Gum Problem*	25
	POW: Linear Nim	15
	The Counters Game	35
	Homework: *Rollin', Rollin', Rollin'*	0

IMP Year 1, The Game of Pig Unit, Teacher Guide
© 2009 Interactive Mathematics Program

6	Discussion: *Rollin', Rollin', Rollin'*	20
	The Theory of Two-Dice Sums	25
	Money, Money, Money	20
	The Theory of Two-Dice Sums (continued)	25
	Homework: *Two-Dice Sums and Products*	0

In the Long Run

7	Discussion: *Two-Dice Sums and Products*	15
	Spinner Give and Take	35
	Pointed Rugs	35
	Homework: *Mia's Cards*	5
8	Discussion: *Mia's Cards*	30
	Reflection: Expected Value and the Unit Problem	5
	A Fair Rug Game?	25
	One-and-One	25
	Homework: *A Sixty Percent Solution*	5
9	Presentations: *POW: Linear Nim*	20
	POW: What's on Back?	20
	Discussion: *A Sixty Percent Solution*	10
	The Theory of One-and-One	40
	Homework: *Streak-Shooting Shelly*	0
10	Discussion: *Streak-Shooting Shelly*	15
	Spins and Draws	35
	Aunt Zena at the Fair	25
	Simulating Zena	15
	Homework: *The Lottery and Insurance—Why Play?*	0
	Homework: *Martian Basketball*	0
11	Discussion: *The Lottery and Insurance—Why Play?*	10
	Discussion: *Martian Basketball*	15
	The Carrier's Payment-Plan Quandary	60
	Homework: *A Fair Deal for the Carrier?*	5
12	Discussion: *A Fair Deal for the Carrier?*	25
	Simulating the Carrier	25

Analyzing a Game of Chance

	The Game of Little Pig	35
	Homework: *Another Carrier Dilemma*	0
	Homework: *Pig Tails*	5
13	Discussion: *Another Carrier Dilemma*	5
	Discussion: *Pig Tails*	15
	Little Pig Strategies	65
	Homework: *Continued Little Pig Investigation*	5
14	Presentations: *POW: What's on Back?*	20
	Discussion and activity: *Continued Little Pig Investigation*	65
	Homework: *Should I Go On?*	5
15	Discussion: *Should I Go On?*	20
	The Best Little Pig	25
	Big Pig Meets Little Pig	35
	Homework: *The Pig and I* and *Beginning Portfolio Selection*	10
16	*Big Pig Meets Little Pig* (continued)	35
	Discussion: *The Pig and I* and *Beginning Portfolio Selection*	15
	"The Game of Pig" Portfolio	30
	Homework: *Take-Home Assessment*	10
17	*In-Class Assessment*	50
	Discussion: *In-Class Assessment* and *Take-Home Assessment*	25
	Unit Reflection	15

Materials and Supplies

All IMP classrooms should have a set of standard supplies and equipment. Students are expected to have materials available for working at home on assignments and at school for classroom work. Lists of these standard supplies are included in the section "Materials and Supplies for the IMP Classroom" in *A Guide to IMP*. There is also a comprehensive list of materials for all units in Year 1.

Listed below are the supplies needed for this unit. General and activity-specific blackline masters are available for presentations on the overhead projector or for student worksheets. The masters are found in the *The Game of Pig* Unit Resources under Blackline Masters.

The Game of Pig

- Dice (at least one pair, different colors, per pair of students)
- Overhead dice
- (Optional) Overhead spinner (clear plastic for making various spinners)
- Cubes of 3 colors (red, blue, and yellow)
- Paper bags (at least 2 bags per group)
- Sentence strips (strips of paper roughly 3 feet by 3 inches) for posting results
- A bag of beans or similar manipulative to serve as counters

More About Supplies:

- Sentence strips are useful in many IMP units. They are often used for posting solutions to problems, for posing problems, and for posting comments, strategies, and questions. These strips can be purchased at educational supply stores or simply made by cutting strips of construction paper, butcher paper, or chart paper.

- Paper bags are often used in probability experiments. Regular lunch bags from a grocery store work fine and are inexpensive and reusable.

- Graph paper is a standard supply for IMP classrooms. Blackline masters of 1-Centimeter Graph Paper, 1/4-Inch Graph Paper, and 1-Inch Graph Paper are provided so that you can make copies and transparencies for your classroom. (You'll find links to these masters in "Materials and Supplies for Year 1" of the Year 1 guide and in the Unit Resources for each unit.)

Assessing Progress

The Game of Pig concludes with two formal unit assessments. In addition, there are many opportunities for more informal, ongoing assessment throughout the unit. For more information about assessment and grading, including general information about the end-of-unit assessments and how to use them, consult the *Year 1: A Guide to IMP* resource.

End-of-Unit Assessments

This unit concludes with in-class and take-home assessments. The in-class assessment is intentionally short so that time pressures will not affect student performance. Students may use graphing calculators and their notes from previous work when they take the assessments.

Ongoing Assessment

Assessment is a component in providing the best possible ongoing instructional program for students. Ongoing assessment includes the daily work of determining how well students understand key ideas and what level of achievement they have attained in acquiring key skills.

Students' written and oral work provides many opportunities for teachers to gather this information. Here are some recommendations of written assignments and oral presentations to monitor especially carefully that will offer insight into student progress.

- *Pig Strategies:* This activity will help you gauge how well students understand the rules of Pig and assess their comfort level with the idea of strategy.

- *0 to 1, or Never to Always:* This activity will illustrate students' grasp of the 0-to-1 scale for probability.

- *Two-Dice Sums and Products:* This activity will show how well students understand and can work with two-dimensional area models.

- *Spinner Give and Take:* This activity can provide a baseline of students' initial understanding of the meaning of "the long run," in preparation for work with expected value.

- *Spins and Draws:* This activity will tell you how well students understand and can work with expected value.

- *A Fair Deal for the Carrier?:* This activity will inform you about students' ability to find probabilities in two-stage situations.

- *Little Pig Strategies:* This activity will tell you how well prepared students are for the detailed analysis of Little Pig.

- The *Best Little Pig:* This activity will inform you of students' grasp of the big picture in the analysis of Little Pig.

Supplemental Activities

The Game of Pig contains a variety of activities at the end of the student pages that you can use to supplement the regular unit material. These activities fall roughly into two categories.

- **Reinforcements** increase students' understanding of and comfort with concepts, techniques, and methods that are discussed in class and are central to the unit.

- **Extensions** allow students to explore ideas beyond those presented in the unit, including generalizations and abstractions of ideas.

The supplemental activities are presented in the teacher's guide and the student book in the approximate sequence in which you might use them. Below are specific recommendations about how each activity might work within the unit. You may wish to use some of these activities, especially the later ones, after the unit is completed. In addition to these activities, you may want to use supplemental activities from *Patterns* that were not assigned or completed.

Average Problems (extension) Three tasks delve into the meaning of the arithmetic average, or mean, of a set of data. Students will likely need to do some numeric experimentation to solve them. These tasks can be used as a follow-up to the discussion of averages that arises during *Waiting for a Double*.

Above and Below the Middle (extension) Students use data from *Waiting for a Double*, a skewed data set, to explore the differences between two common measures of the "middle" of a data set: mean and median.

Mix and Match (reinforcement) This activity is an opportunity for students to work with the basic definition of probability. Students determine the probability of choosing matching gloves from two drawers containing assortments of left-hand and right-hand gloves. The activity requires students to take into account that different outcomes may not be equally likely.

Flipping Tables (extension) Building on work in *What Are the Chances?*, two tasks ask students to explore the difference between the number of combinations possible when flipping coins (when flipping two coins, one combination is one head and one tail) and the number of sequences possible (HT and TH are two different sequences). In the process, students further investigate the important idea of equally likely outcomes.

A Pizza Formula (extension) As a follow-up to *Paula's Pizza,* students build an In-Out table to investigate the relationship between the number of available toppings and the number of two-topping pizzas that can be made.

Heads or Tails? (extension) This activity is another opportunity for students to use a rug diagram to analyze two-dice sums. Given the probabilities for the various outcomes of flipping two coins, students investigate the probabilities for the outcomes of flipping three, four, and ten coins. You may want to allow up to a week for students to work on this activity.

Small Nim (reinforcement) This activity explores simpler cases of the game Linear Nim and can help students get started with that POW. Students evaluate several strategies for playing the game using versions with only four or five marks.

Piling Up Apples (extension) In this strategy activity, students develop and evaluate strategies for winning a nonprobabilistic game similar to Linear Nim. You may want to assign it after the presentations on the POW Linear Nim.

Counters Revealed (extension) Students analyze a simplified version of the game presented in *The Counters Game*.

Two-Spinner Sums (extension or reinforcement) Beginning in *Spinner Give and Take,* students use spinners to model probabilistic situations. In this set of two tasks, they examine the results of spinning two spinners and compare this work to the results of tossing two dice.

Different Dice (reinforcement) This activity, in which students investigate the probabilities for various sums and products of two modified dice, is similar to *Two-Dice Sums and Products* and brings out the notion that the sum of all the probabilities in a given situation must equal 1. You might use it with students who need more practice with ideas like those in that activity.

Three-Dice Sums (extension or reinforcement) In this activity, students find probabilities for the outcomes of rolling three dice.

Make a Game (extension) Students create, test, and write an account of their own games that involve probability and strategy.

Pointed Rug Expectations (reinforcement) In this follow-up to *Pointed Rugs,* students find the long-run average results, or expected values, for a variety of rug diagrams.

Explaining Doubles (extension) Students who have developed the tools of theoretical probability return to *Waiting for a Double* and explain the expected values found in that activity.

Two Strange Dice (reinforcement or extension) As in the payment-plan activities, students compare a constant point value to the expected value of a probabilistic situation—this time, in rolls of a pair of nonstandard dice.

Expected Conjectures (extension) This activity focuses on the idea that the answer obtained by using the "large number of trials" method is actually independent of the number of trials. It also suggests the idea of multiplying each possible numeric outcome by its probability and adding the results to get the expected value. You will want to allow substantial time for students to work on this activity.

Squaring the Die (extension) In another activity that offers an opportunity to gain deeper insight into expected value, students explore how expected value is affected when events are combined. You will want to allow substantial time for students to work on this activity.

Fair Spinners (reinforcement) Question 1 of this activity is similar to A Fair Rug Game?, using the spinner from *Spinner Give and Take* instead of a rug, and is also similar to Question 1 of Spins and Draws. Question 2 of this activity offers students an opportunity to create their own spinner games.

Free Throw Sammy (reinforcement) In this follow-up to *One-and-One,* students construct an area model for a success rate of 80 percent and use it to determine probabilities for each outcome and the expected value.

Which One When? (extension) As a follow-up to *The Theory of One-and-One,* this activity brings out the idea that knowing the expected value for a situation does not always give all the information needed to make a decision.

A Fair Dice Game? (reinforcement or extension) Building on *A Fair Rug Game?,* this activity requires students to apply their knowledge of two-dice sums as well as the concept of a fair game.

Get a Head! (extension) This activity asks students to apply their developing understanding of probability and expected value to problems involving repeated flips of a coin. Question 2 asks them to analyze a game that could be arbitrarily long.

More Martian Basketball (extension) In this follow-up to *Martian Basketball,* students compare expected outcomes for three free-throw shooters. The choices might depend on how many points the team is ahead or behind.

Interruptions (extension) Students might employ area models, tree diagrams, and simulations as tools for analyzing the situation in this activity, though a general solution is too complex to expect at this stage. You may want to allow up to a week for students to work on it.

Paying the Carrier (extension) In this follow-up to the *The Carrier's Payment-Plan Quandary* and related activities, students are asked to devise a different payment system for the carrier. It essentially requires students to develop and solve a linear equation and then generalize the result.

Pig Tails Decision (extension) This follow-up to *Pig Tails* poses a strategy question for the game of Pig Tails in preparation for similar questions about Little Pig and Big Pig that students will soon encounter.

Pig Strategies by Algebra (extension) As an extension of *Should I Go On?* and *Big Pig Meets Little Pig,* students are asked to state strategies for Little Pig and Big Pig algebraically, with point values expressed as variables.

Fast Pig (extension) This activity uses a variant of the game of Pig to lead students toward a general analysis of the expected value for an n-roll strategy for Pig. You may want to allow up to a week for students to work on this activity.

Chance and Strategy

Intent

The activities in *Chance and Strategy* are designed to accomplish two goals. First, students are introduced to the game of Pig and have opportunities to begin to develop strategies for playing the game. Second, students encounter several situations that challenge their intuitive notions of probability.

Mathematics

The six possible rolls of a fair die are equally likely to occur. In Pig, a dice game, rolling 2 through 6 will earn a player points. Players may keep rolling until they decide to stop or until they roll 1. Rolling 1 will end the turn, and the score for that turn will be 0. What is the best strategy for playing this game? In other words, at what point should a player stop rolling in order to earn the most points possible for that turn? In *Chance and Strategy*, students begin to test and articulate strategies for optimizing their turns. The idea of a complete strategy—one that can be applied in all possible game situations—is introduced.

Probabilistic reasoning frequently runs counter to intuition. Students often need a great deal of experience before they become willing to base their predictions on probabilistic notions rather than on such ideas as "being lucky" or that a particular result is "due" to occur.

These activities give students some concrete experiences, using dice and coins, with probabilistic questions. The two big ideas underlying these experiences are **independence** (each flip of a coin, for example, is unaffected by previous flips) and **random** variability (it is likely that the result of flipping 50 coins will not be exactly 25 heads).

Students are also introduced to the idea that the probability of an event can be quantified using real numbers between 0 (impossible) and 1 (certain) by counting the occurrences of that event and comparing that to the total number of events possible. The POW introduced in *Chance and Strategy,* in addition to being another rich problem to solve and to write about, gives students the opportunity to develop systematic counting methods.

There is growing research supporting the idea that technical vocabulary is best learned by students when they attach the terms to concepts they already understand—meaning that it might be advantageous to introduce terminology as students' work on important ideas warrants. To illustrate, consider the following terms in the context of the activity *Waiting for a Double*:

- The *experiment* is rolling a pair of dice.

- The possible **outcomes** of this experiment include rolling 2 on the first die and 5 on the second, or 1 on the first die and 4 on the second.

- The **event** of interest is rolling a double (such as 3 and 3, or 6 and 6).

- The **probability** of rolling a double is the ratio of the number of ways this event can occur to the total number of outcomes of this experiment because each possible outcome is equally likely.

Progression

In the opening class and homework activities, students play the game of Pig. They will return to the game later in the unit, once they have developed tools to analyze playing strategies. Subsequent activities in *Chance and Strategy* begin to develop these tools. The first POW of the unit is also assigned.

The Game of Pig

POW 3: A Sticky Gum Problem

Pig at Home

Pig Strategies

Waiting for a Double

The Gambler's Fallacy

Expecting the Unexpected

Coincidence or Causation?

What Are the Chances?

Paula's Pizza

0 to 1, or Never to Always

The Game of Pig

Intent
This activity introduces students to the game of Pig and to the study of probability. Students develop, test, and compare strategies for playing the game as they begin to consider the central problem of the unit.

Mathematics
Students will collect data about rolling a die as part of playing the game of Pig. After some experience with the game, they will be able to articulate that a **strategy** is a complete plan of action intended to reach a particular goal.

Progression
Students are introduced to the game of Pig.

Approximate Time
35 minutes

Classroom Organization
Groups of 4, then pairs, followed by whole-class discussion

Materials
Dice

Doing the Activity
After students have read *The Game of Pig* and, in particular, "Your Assignment," ask a few students to demonstrate how to play the game and how to keep score. Play a few turns as a class to clarify the rules. Be sure students understand the distinction between a *single roll of the die* and a *turn*. Each student should keep a record of the single rolls and the score for each turn. These records will allow students to review how the game progressed and to analyze the outcome of each turn.

After students play several turns in groups of four, have them play a game or two in pairs (as time permits) and experiment with playing strategies.

Discussing and Debriefing the Activity
Ask students to discuss some of the playing strategies they used. For a given turn, a basic strategy defines when to roll again to try to earn more points and when to stop rolling to not risk losing the points already earned that turn. You might ask each group to post at least one of their strategies. When duplicate strategies appear, request alternate strategies.

After posing the Key Questions, emphasize that in mathematics a strategy is a complete plan of action intended to reach a particular goal. A strategy for playing Pig must tell whether or not to roll again in every circumstance in which there is a choice.

Key Questions

What do you think a strategy is?

How do you use strategies in your daily life?

What were some high scores for individual turns?

How did you decide whether to roll again or to stop?

What do we mean by the "best" strategy?

How did your strategy change as you played Pig and understood the game better?

Does this group's strategy seem complete and clear?

How can we compare different strategies?

How might you measure whether one strategy is any better than another?

Can another group test your strategy?

POW 3: A Sticky Gum Problem

Intent
As with all POWs, students explore a problem outside of class and communicate their thinking in writing. This POW taps students' developing skill at finding and using patterns from the previous unit, and asks students to count possible outcomes, a strategy used throughout the current unit to find probabilities. This POW also helps students understand what is meant by a generalization.

Mathematics
At the heart of this activity is the counting concept known as the *pigeonhole principle*. Informally, the pigeonhole principle states that if you have more pigeons than pigeonholes, at least one of the holes will contain more than one pigeon. In this activity, the "pigeonholes" are different colors of gum balls. If two children want to buy gum balls of the same color from a machine containing gum balls of three colors, *in the worst case*, the first three gum balls would be three different colors (they will fill the three pigeonholes). The next gum ball would have to match one of the first three. So, the children would have to buy *at most* four gum balls to be sure to get two of the same color.

Progression
This POW is introduced at the start of the unit, with presentations about a week later.

Approximate Time
10 minutes to pose activity and short periods to check on progress

1 to 3 hours for activity (at home)

20 minutes for presentations

Classroom Organization
Groups of 3 or 4 for brief segments, concluding with whole-class presentations

Doing the Activity
Students can begin working on this POW on the first day of the unit. The activity begins with several special cases and then asks students to search for a general method, given any number of children and any number of colors, for finding the number of gum balls one must buy to be sure the children all get the same color gum ball. A class discussion of the first special case—in which the answer is given and students are asked to explain why it is correct—might help students get started.

Discussing and Debriefing the Activity

After a week or so of having students work outside of class, with perhaps a portion of one day devoted to a progress check, have some students present their results.

Focus the discussion on the process of trying different examples, organizing the information, looking for patterns, expressing patterns symbolically, and explaining patterns, rather than on specific formulas. At this early stage, you might expect only some students to arrive at a general formula.

Key Questions

In each specific case, what is the smallest number of gum balls the children could purchase and all get the same color? Why?

In each specific case, how many gum ball purchases would guarantee that the children would all get the same color? Why?

For any specific number of children, how does the number of colors available affect the number of gum balls one would need to buy? What patterns can help you to answer this question?

For any number of available colors, how does the number of children affect the number of gum balls one would need to buy? What patterns can help you answer this question?

Pig at Home

Intent
In this homework activity, students become more comfortable with the game of Pig and have another opportunity to verbalize, test, and explain playing strategies.

Mathematics
Students continue to collect die-rolling data and to discuss and compare strategies.

Progression
Exploring at home, students spend more time thinking about and developing clear strategies for playing Pig that, in the long run, will produce the highest possible score.

Approximate Time
5 minutes for introduction

30 minutes for activity (at home)

15 minutes for discussion

Classroom Organization
Individuals, followed by whole-class discussion

Materials
Dice

Doing the Activity
To introduce the activity, read *Pig at Home* out loud as a class. Have students agree that they should continue to keep track of both the individual die rolls and the score earned on every turn.

Discussing and Debriefing the Activity
Students' strategies will be discussed in the next activity.

Pig Strategies

Intent
In this activity, students focus on the need for clarity when communicating a **strategy**—that is, a complete plan of action.

Mathematics
In this unit, the "best" strategy is the one that produces the highest average score per turn in the long run. Later in the unit, students will learn the formal name for this concept, **expected value**. For now, you can begin to focus their attention on the idea of an average score per turn.

Progression
Through their work in groups and with the whole class, students will come to understand the need to articulate strategies clearly and completely.

Approximate Time
10 minutes for small-group discussion

20 minutes for whole-class discussion

Classroom Organization
Small groups, followed by whole-class discussion

Materials
Sentence strips

Doing the Activity
Ask all students to report their homework results from *Pig at Home* to their groups. This is an opportunity to reinforce a productive group dynamic and to emphasize the importance of all students doing their homework.

Focus students' constructive critiques on the clarity and completeness of each strategy. For example, a student might say, "Your strategy doesn't tell me what to do if . . . ."

Ask each group to decide which of the strategies reported is the best one the group has discovered so far. Have each group write this strategy on a sentence strip and post it for reference during the class discussion.

Discussing and Debriefing the Activity

When all the strategies are posted, ask groups to test some of them. For example, groups could compare the total scores after ten turns using each of several strategies.

You might also ask students to imagine two players using two different strategies. One player has taken 50 turns and scored 322 points; the other has taken 60 turns and scored 348 points. Ask each group to try to come to a consensus about which player seems to have used the better strategy. Let some volunteers present their groups' decisions and reasoning.

As needed, bring out the idea that at the rates the players are going, when each player has taken 100 turns (or any other particular number), the first player will have more points than the second because the first player has a higher *average score per turn*.

In this context, "best" means "the most points per turn in the long run." You may want to point out explicitly that the best strategy might not necessarily give the highest score in a particular game. The best strategy is the one that is most likely to help a player win in the long run.

Students' central task in this unit will be to find the best strategy for playing Pig. Post this goal for reference during the rest of the unit:

> **Unit Goal**
> To find the strategy for playing Pig that gives the most points per turn in the long run.

Key Questions

Can you define a "best" strategy?

How can we describe the highest average score per turn in the long run?

Does this group's strategy seem complete and clear?

What do we mean by the "best" strategy?

Which of these two players seems to have used the better strategy?

Waiting for a Double

Intent
In this activity, students conduct their first simulation, collecting, displaying, and summarizing data to test a conjecture about probability. The activity draws upon their intuitive sense of probability and engages them in thinking more precisely about the topic. This is one of several activities in which intuitive notions about probability can be at odds with experimental or theoretical analyses.

Mathematics
When rolling two dice, what is the average number of rolls needed to roll a double? In this activity, students approach this question through collecting and analyzing data. By rolling two dice repeatedly, they collect many examples of how many rolls it can take to get a double. It could take a single roll, or it could take 30 rolls or more. (The distribution of their results will have an average value of about 6.)

Progression
Up to now, students have been exploring the unit problem informally. Now they begin a series of explorations designed to help them develop the tools they will need to answer the unit problem. The in-class portion of this activity has several parts: reviewing the homework, gathering class data, and graphing the class data.

Approximate Time
5 minutes for introduction

20 minutes for activity (at home or in class)

25 minutes to collect and analyze class data

Classroom Organization
Individuals, then small groups to review data, then whole class to gather and analyze data

Materials
Pairs of dice

Doing the Activity
Students can do this activity for homework. When assigning the activity, you might have them try a few trials to be sure they know what they are counting and to consider the range of possible outcomes.

Students are first asked to predict the average number of rolls it will take to get a double. Some students might have ideas about this based on experience with dice, while others might simply guess. Then they are to collect the outcomes of ten experiments. Stress that it is important to record the results exactly as they occur.

Finally, students are asked to review their data, compute the average number of rolls for the ten experiments, and compare this average to their initial predictions. Through this process, students connect the average outcome of repeated trials with the expected number of rolls.

Discussing and Debriefing the Activity

In small groups, have students compare their predictions with the results of their data collection. It is likely that many students will have at least one value as large as 15 rolls, and some may have values as large as 30 rolls or more.

As a class, the ultimate goal is to create a display of the data collected by all of the students. Begin by giving students a chance to report to the class their initial predictions, how their experiments turned out (that is, the fewest and greatest number of rolls it took to get a double, and the average of their ten trials), and what they now think about their predictions.

Next, ask the class how they might display the collected data. Students may have several ideas for graphing the outcomes. As a class, decide to make a **frequency bar graph** where each bar would represent a different number of rolls needed to get a double. The height of each bar would show the number of times that result occurred.

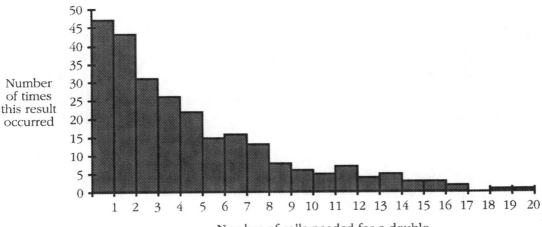

Once the class data have been graphed, ask students to make observations. Emphasize what a frequency bar graph can reveal about data, and refresh students' memories about the concepts of **mode** and **median**.

Ask students to compute the class average for the number of rolls needed to get a double. This value should be close to 6. Some students may be surprised to see far more results below 6 than above 6. Others may note that there are far more bars above 6 than below 6. As needed, help them understand that a single large result "balances" lots of small results, and that in general the average of all the frequencies does not equal the average of the actual results.

Finally, ask students to review how their individual averages varied and how those averages compare to the class average.

If you have access to the technology you might demonstrate or have students explore simulations. Class data can be simulated using dynamic data software such as Fathom Dynamic Data. Or you might use a graphing calculator.

Key Questions

Before assigning the activity:

How would you do this experiment?

Why is it important to get genuine data when doing activities like this?

How do you compute the average of a set of numbers?

During the class discussion:

How did you make your prediction?

What were the highest and lowest number of rolls you got?

How might you graph one student's outcomes?

Can you visualize the average in terms of the graph?

How might we graph the outcomes of the entire class?

Is 1 a likely outcome?

Supplemental Activities

Average Problems (extension) has three tasks that delve into the meaning of the arithmetic average, or mean, of a set of data.

Above and Below the Middle (extension) offers further experience with mean and introduces the concept of median.

The Gambler's Fallacy

Intent
Students come into class with a limited set of experiences with which to think about probability. In this activity, students gather data to explore, in an informal way, a commonly held misconception about the important ideas of randomness and independent events. In the process, they develop tools to analyze probabilistic situations: namely, simulations and the concept of experimental probability.

Mathematics
When you flip a fair coin, you are equally likely to get heads or tails. When you flip a coin repeatedly, heads or tails will come up randomly. If you get three heads in a row, what is the chance that the next flip will be a head? *The Gambler's Fallacy* refers to the idea that the next flip is more likely to be a tail because getting four heads in a row is very unlikely. In fact, each flip of the coin is **independent**. A coin has no memory, and the chance of a head is always the same as the chance of a tail.

Progression
This is the second of a number of activities in which students gather data to investigate a probability question. In this activity, students toss a coin many times to investigate what happens after a run of heads or tails. Students pool their data to try to assess predictions that are based on the gambler's fallacy.

Approximate Time
20 minutes for data collection

30 minutes for discussion

Classroom Organization
Pairs to collect data, followed by whole-class discussion

Materials
Coins

Cardstock or index cards

Doing the Activity
When students read the introduction to *The Gambler's Fallacy,* three points of view generally emerge regarding the roulette strategy of betting on black after a string of reds:

A string of reds makes the next spin more likely to be red (as if a trend is happening).

A string of reds makes the next spin more likely to be black (as if the string of reds must be compensated for).

Previous spins have no effect on the present spin.

Take a straw poll to see how many students agree with each of these ideas.

Have students read "The Experiment." Then ask how this experiment is related to the roulette strategy in which gamblers expect a change in result after a string of a given outcome.

Have students work in pairs to begin the experiment and record their results. Most pairs will need to take some time to devise a method for counting "sames" and "differents." Here is one technique you might suggest. Given this series of coin flips:

T H T T T T H T T

Students can cut a rectangular "window" from an index card or cardstock and move it systematically over the sequence of outcomes, looking for triplets, and giving results like this (a "same"):

T H [T T T T] H T T

and this (a "different"):

T H T [T T T H] T T

Once pairs have compiled the results of their experiments, total the number of "sames" and "differents" for the entire class.

Discussing and Debriefing the Activity

It is unlikely that the number of "sames" will be equal to the number of "differents," and there is a chance that these numbers will differ enough to be seen by students to support the gambler's fallacy. With more data, the number of "sames" will get closer to the number of "differents."

The gambler's fallacy experiments can lead naturally into defining and discussing **theoretical probability**, **observed probability**, and **independent events**. This experiment produces data that students can use to determine the *observed* (or *empirical*) probability that the toss after a triplet is the same as the letters in the triplet. The gambler's fallacy is a fallacy because, according to *theory*, every flip of a coin has an equal chance of coming up heads as tails, regardless of what has happened before. In other words, each toss of a coin is *independent* of previous flips.

Flipping a fair coin is a **random** experiment. All of the outcomes of this process—a head or a tail—are equally likely to occur. However, equally likely outcomes do not necessarily result in equal numbers for each outcome when the experiment is done. For example, flipping a coin 100 times is unlikely to result in exactly 50 heads and

50 tails. However, as this experiment is done more and more times, the number of heads will get closer and closer to half the total number of tosses.

You can connect the gambler's fallacy with the central unit problem by asking, **How is the gambler's fallacy related to strategies for the game of Pig?** Specifically, ask whether students thought that the chance of rolling 1 became greater if they had already rolled something other than 1 several times.

It's not reasonable to expect to dispel the misconception of the gambler's fallacy entirely, even if students state unequivocally that they would not be so foolish as to accept it. Even knowledgeable adults operate on the basis of this idea at least occasionally. Sometimes just giving a name to such a mistaken idea can help people identify and avoid it.

You might use Fathom Surveys to pool the data from several classes. With Fathom Dynamic Data™ software you can also simulate data for one or many classes.

Key Questions

What do you think about the strategy described in the activity?

Why does this activity talk about a roulette wheel but tell us to use coins?

Why are the number of "sames" and the number of "differents" so close?

Why doesn't what has happened already affect the next flip of a coin?

How is the gambler's fallacy related to the roulette strategy?

How is the gambler's fallacy related to strategies for the game of Pig?

Expecting the Unexpected

Intent
This activity gives students experience with experimental variation as they continue to develop the skills of collecting and organizing data. Students also make predictions and compare their predictions with their observations.

Mathematics
Gathering data about coin flips will help students get a sense of experimental variation. Students combine data generated by the class and make a frequency bar graph of the results, practicing the skill of keeping track of experimentally generated data. They also examine different ways to group data to make a frequency bar graph.

Progression
This activity engages students' intuitive ideas about probability. Later in this unit, students will consider random events and equally likely outcomes more closely.

Approximate Time
5 minutes for introduction

30 minutes for activity (at home or in class)

15 minutes for discussion

Classroom Organization
Individuals, then groups, followed by whole-class discussion

Materials
Coins

Doing the Activity
To introduce the activity, ask, **How many heads would you expect to get if you flipped a coin 50 times?** Encourage discussion. It is likely that some students will immediately respond with 25 and that others will suggest a range. You might also have students share their initial predictions for Questions 2a and 2b.

This activity will likely be done as homework. Remind students that the discussion at the next class meeting will depend on all students bringing in the results of their experiments.

Discussing and Debriefing the Activity
Have students share their data with the class. Students might be surprised that the most likely outcome—25 heads out of 50 flips—is not particularly likely, happening only about 11% of the time. If you didn't bring up the "most likely versus likely"

distinction earlier in the unit, this is a good time to do so. (In a class of 30 students, with the 50-flip experiment done a total of 60 times, the highest result would be approximately 33 heads, and about 7 of the 60 experiments would result in exactly 25 heads.)

Ask, **How might we graph the data?**

Students may have several ideas about graphing the data. This is a good time to pursue the idea of grouping data using a frequency bar graph, which the class encountered in *Waiting for a Double*.

To encourage students to begin thinking about how to organize the data so that each student's data are included, ask, **What planning will we need to do as a class so that each of us can have a frequency bar graph that shows all our data?** Depending on students' level of autonomy, and what you wish to challenge them with, consider how much input into the organization of this data collection and representation you wish to have.

Ask, **How might you group the data?**

Have each group choose its own way of grouping the results and making the corresponding frequency bar graph. Display some of the bar graphs that groups create. As an example, students may produce a graph something like this.

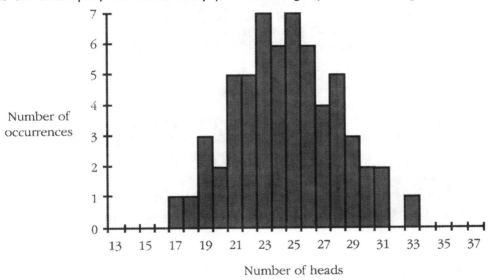

You might ask if the graph can be redrawn with fewer bars to make the "big picture" more obvious. **How might you modify the graph to see the "big picture" better?** You can use the term *bin width* to describe the size of each grouping section. For instance, a bar representing results from 23 through 26 has a bin width of 4.

If someone suggests making a graph in which the bin widths are not all the same, you might ask whether this organization could be misleading. For example, if results 27 and over were combined in one bar and all other results represented as single cases, the graph would seem to indicate that the most likely outcome is a result of at least 27. For this reason, we typically make all groups the same "width."

For example, the graphs below combine frequencies for two outcomes at a time (bin width = 2) and three outcomes at a time (bin width = 3).

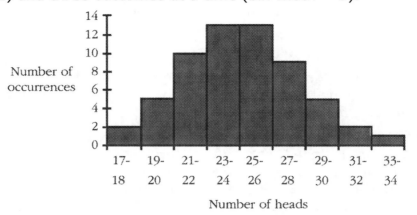

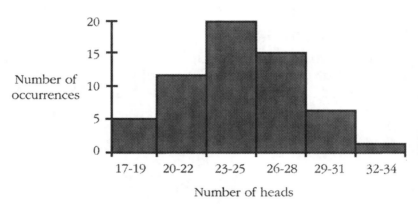

Students may debate the merits of various grouping methods. For instance, some may argue that if they are combining three bars at a time, they should work symmetrically around the value 25, so that the middle group is 24–26.

To close the activity, summarize how some of students' initial ideas about what is most likely to occur when flipping 50 coins seem to appear within the frequency bar graph. You may want to point out that, generally speaking, all of these graphs are higher in the middle and lower at the ends, whereas the frequency bar graph of the data from *Waiting for a Double* started out high and gradually got lower.

If you have the necessary technology you can show or have students explore a Fathom Dynamic Data™ software simulation of this activity.

Key Questions

How many heads would you expect if you flipped a coin 50 times?

What planning will we need to do as a class so that each of us can have a frequency bar graph that shows all our data?

How might we graph the data?

How might we group the data?

How might we modify the graph to see the "big picture" better?

Coincidence or Causation?

Intent
In this activity, students consider whether previous experiences should be taken into account in determining the probability of a particular event. This activity will give students further insight into the gambler's fallacy.

Mathematics
This activity focuses informally on the concept of *independence*, asking students to make judgments about whether past occurrences affect future events.

Progression
This activity continues the informal development of probability, drawing on students' intuitive notions. The idea of independence arises and is connected to the gambler's fallacy.

Approximate Time
5 minutes for introduction

15 minutes for activity (at home or in class)

15 minutes for discussion

Classroom Organization
Individuals, followed by whole-class discussion

Doing the Activity
This activity follows upon the discussion and defining of independence in *The Gambler's Fallacy*. Introduce the activity by mentioning that there are times when an occurrence *does* affect a future event. You might suggest examples of your own or encourage students to describe some they know.

Read the introduction to this activity as a class. Clarify that students are to determine whether, in each situation, the past will influence the future and to write a paragraph explaining their answer.

Discussing and Debriefing the Activity
Spend a few minutes discussing each of the three situations. In each case, there are reasonable arguments on both sides. Again, the goal of this activity is to clarify the ideas of independence and dependence, rather than to resolve these specific problems.

Here are some plausible arguments that the new situation *does not* follow the general probabilities.

The baseball player's chances may be greater than 1 in 7 because he is in especially good condition, because he is playing against a team with poor pitchers, or because he is playing in a small park for several games in a row.

Your chances of getting a red light may be more than .5 because the lights along Pine Street are not synchronized or because the lights are on a timer and Mr. Bryant always leaves the house at exactly the same time.

If the store where you bought the cones is particularly careless in handling merchandise, the chances of finding a broken cone may be more than 1 in 100. It's also possible that Happy Days exaggerates their quality control.

Review the term **independent events**, introduced in *The Gambler's Fallacy*, to describe events whose probabilities do not depend on each other. Help students to see that the questions in this activity can be thought of as asking whether the situations involve independent events.

Key Question
Which of the situations involve independent events?

What Are the Chances?

Intent
The gambler's fallacy experiments lead into discussions of theoretical and observed probabilities and of independent events. In this activity, students begin thinking numerically about probability, using the scale for probabilities of 0 to 1 and distinguishing between theoretical and observed probability.

Mathematics
The most important idea in this activity is that representing the probability of an event as $\frac{\text{the number of outcomes you're interested in}}{\text{the total number of possible outcomes}}$ is accurate only when all of these **outcomes** are *equally likely*. This definition of probability also helps students see why a probability must take on a value between 0 and 1, inclusive. Probabilities can be expressed as fractions, decimals, and percentages; students will be using all three forms in this unit.

Progression
This activity begins with the teacher introducing the fraction definition of probability, focusing on equally likely outcomes.

Approximate Time
35 minutes for introducing and doing the activity

30 minutes for discussion

Classroom Organization
Groups, followed by whole-class discussion

Materials
Sentence strips

Doing the Activity

Part I: Finding Probabilities
The focus in this activity is on building intuition. For some problems, a reasonable estimate should suffice, even if the problem has an exact theoretical answer. Encourage students to think in terms of both observed and theoretical probabilities.

Begin by asking students, **What are some examples of events that are impossible? Some examples of events that are certain?** Keep a list of their suggestions. Then ask for examples of events whose chances of occurring are between "impossible" and "certain," and list those as well.

Next, draw a segment of a number line from 0 to 1.

Explain that mathematicians place "impossible" events at the left end of this number line and "certain" events at the right end. Then ask students to come up and indicate where they think some "between" events from the list should be placed.

Introduce the "P(...) = " notation for stating the probability of an event. For example, when a fair coin is flipped, it should come up tails half the time in the long run. This probability is written

$$P(\text{tails}) = \frac{1}{2}$$

Point out that people often think of an activity or situation as potentially having one of several results. Introduce the term **outcomes** for describing these possible results. Ask, **What are the possible outcomes for flipping a coin?** (heads or tails) **For rolling a die?** (1, 2, 3, 4, 5, or 6) **For handedness?** (left-handed, right-handed, or ambidextrous)

Ask the class how they could calculate probability when an event has only certain outcomes that can occur and they make the assumption that each outcome is equally likely. **How can you calculate the probability of an event when the outcomes are equally likely?** For example, one generally assumes that the two possible outcomes of a coin flip, and the six possible outcomes of a die roll, are equally likely. You might mention that probabilities based on such assumptions are examples of **theoretical probabilities**.

Students should see that the probability of any one of these outcome is

$$\frac{1}{\text{the total number of outcomes}}$$

For example, for a die roll, there are six possible outcomes, each equally likely. So $P(\text{rolling } 5) = \frac{1}{6}$.

We are often interested in a particular *set of outcomes* of some larger set. For example, out of the six possible outcomes for a die roll, we might want to know the probability of rolling either 1 or 2. Elicit from the class the idea that, in general, the probability of getting a result that is within a specific set of outcomes you're interested in is expressed by the fraction

You can illustrate with examples. For instance, ask, **What is the probability of rolling 1 or 2 with a standard die?** Students should be able to explain that there are six possible outcomes, and that since each is equally likely,

$$P(\text{rolling 1 or 2}) = \frac{2}{6}$$

Acknowledge that it is not always clear whether the possible outcomes in a given situation are equally likely. Of the three examples given earlier—flipping a coin, rolling a die, and handedness—students will presumably know that the outcomes for handedness are not equally likely. Consider the question, **What is the probability that a person is right-handed?** Although there are three possible outcomes, the probability that a person is right-handed is much greater than $\frac{1}{3}$.

Students may assert that even the outcomes for coins and dice are not necessarily equally likely. Explain that, unless otherwise indicated, they should assume that coins and dice are fair—that is, all outcomes from a coin toss or a die roll are equally likely.

Another example you might use is, **What is the probability that someone was born in (your state)?** Again, students should realize that the various possibilities are not equally likely.

Students should recognize that any fraction of the type

$$\frac{\text{number of outcomes you're interested in}}{\text{total number of possible outcomes}}$$

must be between 0 and 1, because the numerator cannot be negative and cannot be larger than the denominator.

Bring out the idea that if the set of "outcomes you're interested in" contains all possible outcomes, you are certain to get one of the results you are interested in, and the fraction is thus equal to 1. Similarly, if there are no "outcomes you're interested in," the probability is 0.

As you work with various examples, illustrate that probabilities can also be written as decimals and percentages. For example, students can write "P(tails) = .5" or "P(tails) = 50%" instead of "P(tails) = $\frac{1}{2}$".

Fractions, decimals, and percentages are all common ways to talk about probabilities, although the definition given above makes fractions the most natural form for the initial discussion.

Part II: Probabilities on the Number Line

One nice way to set up this activity is to give groups identical strips of paper showing a number line marked from 0 to 1. Groups can write a large letter on their number lines to indicate the probability for each outcome. If the number lines are then displayed during the class discussion one directly below the other, students can see whether the letters line up.

As you circulate among groups, you may see that a common misconception is that all outcomes in a situation are equally likely. This misconception often involves the way in which individual outcomes are identified.

For instance, in question D, students may view the situation as involving three outcomes—flipping two heads, two tails, and one of each—and incorrectly conclude that each outcome has a probability of $\frac{1}{3}$.

You may wish to clarify this by returning to example A and arguing that because there are three colors, the probability of picking a blue gum ball is $\frac{1}{3}$. Students will likely see that the colors are not equally likely to be chosen. If needed, help them recognize that it is more useful to think of each *gum ball* as a possible outcome rather than each *color*. Thus, they can think of the situation as having nine possible outcomes rather than three and see that the probability of getting a blue gum ball is $\frac{2}{9}$.

Returning to Question D, ask groups to find a way to express the problem in terms of equally likely outcomes.

Discussing and Debriefing the Activity

You might begin by having groups display their number lines one above the other, so everyone can see how well the letters match up from group to group.

Help students to distinguish between probabilities based on abstract models and probabilities based on observed or experimental results. For the latter, the probabilities are open to considerable interpretation. For example, you might ask, **Are you basing your answer on observed data or on intuition? For example, has it ever snowed in Florida in July?**

Although there may be more disagreement on the problems involving observed probability, make sure students understand and can explain the mathematics in the simpler examples involving theoretical probability.

Question A is an opportunity for the class to articulate again the definition of probability as

$$\frac{\text{number of outcomes you're interested in}}{\text{total number of possible outcomes}}$$

Emphasize that this formula applies only to situations with equally likely outcomes.

Question B and C are based on observation, such as information gained from reading and weather reports.

For Question D, it is unlikely all students will see why this example has a probability of $\frac{1}{2}$. You might ask students to list the possible outcomes. The key is realizing that getting a tail and then a head is different from getting a head and then a tail. In other words, the equally likely outcomes are HH, HT, TH, and TT. Since the two flips are different in two of the four cases, the probability of getting different results is $\frac{2}{4}$.

For Question F, students may suggest that the teacher's choices aren't random. In this case, the problem concerns observed probability, and all sorts of answers are possible. If the two students *are* selected at random, the probability depends on the size of the class.

For Question H, at least a few students will likely see from a theoretical model that the probability is $\frac{1}{6}$. One approach is to imagine rolling one die first; in that case, the second die has a $\frac{1}{6}$ chance of matching the first. Another approach is to list all 36 two-die pairs (identifying them as equally likely), and note that 6 of these pairs are doubles. You may want to point out that the answer to Question H seems consistent with the result in *Waiting for a Double*, in which it takes an average of about six rolls to get a double.

For Question I, students could use their experimental data from *Waiting for a Double* to estimate the probability. (Don't expect students to be able to figure out the theoretical value, which is $\frac{91}{216}$, or about 42%.)

Key Questions

What are some examples of events that are impossible? Events that are certain? Events between impossible and certain?

Supplemental Activities

Mix and Match (reinforcement) asks students to determine the probability of choosing matching gloves from two drawers containing assortments of left-hand and right-hand gloves. The activity requires students to take into account that different outcomes may not be equally likely.

Flipping Tables (extension) builds on the ideas in this activity and defines the terms combination and **sequence**.

Paula's Pizza

Intent
In this activity, students make a list of all possible cases, giving them a clear picture for finding probabilities.

Mathematics
Using systematic lists, students can find probabilities by counting possible outcomes and applying the basic definition of probability:

$$\frac{\text{number of outcomes you're interested in}}{\text{total number of possible outcomes}}$$

Progression
Students are introduced to another tool for finding probabilities.

Approximate Time
Brief introduction

20 minutes for activity (at home or in class)

15 minutes for discussion

Classroom Organization
Individuals, followed by whole-class discussion

Doing the Activity
As a class, read the introduction to *Paula's Pizza* aloud. Clarify that all pizzas must have two different toppings; no pizza can be "double" pepperoni or "double" anything.

Students will probably want to make a list of all possible two-topping combinations. Encourage them to be systematic, perhaps listing all the combinations with sausage, then those with onions (and not sausage), and so on. There are several ways to organize this list to ensure that no two-topping pizza is overlooked and that there are no duplicates.

Discussing and Debriefing the Activity
Encourage students to compare their results. Did everyone find all the possible combinations, without duplication? With the list of 15 possibilities, students should realize that Paula's probability of getting the pizza she ordered (Question 2) is $\frac{1}{15}$ and that her probability of getting something different is $\frac{14}{15}$. If students do not

notice that these two probabilities add to 1, ask questions to help lead them to this observation.

In Question 3, students just need to count that there are six combinations that include neither sausage nor pepperoni. They can then conclude that the probability of Paula getting a pizza she likes is $\frac{6}{15}$ and the probability of her getting a pizza she doesn't like is $\frac{9}{15}$.

Key Questions

Can you list all the outcomes?

What do you notice about the two probabilities in Question 2?

Why is the sum of the probability of Paula not getting what she ordered and the probability of Paula getting what she ordered equal to 1?

Supplemental Activity

A Pizza Formula (extension) challenges students to find a proof of the pattern and a **closed formula**.

0 to 1, or Never to Always

Intent
In this activity, students work with probabilities expressed as fractions, percentages, and decimals.

Mathematics
Students continue to focus on the distinction between probability based on a theoretical model and probability based on observed results and on measuring probability using a real-number scale from 0 to 1.

Progression
Earlier in the unit, students estimated the probability of given events. In this activity, they invent situations that match a given probability.

Approximate Time
5 minutes for introduction

25 minutes for activity (at home or in class)

10 minutes for class discussion

Classroom Organization
Individuals or pairs, followed by whole-class discussions

Doing the Activity
Clarify that students are to invent two situations for each question, one based on theoretical probability and the other on observed results.

Discussing and Debriefing the Activity
If students have worked individually, ensure that there is time for them to compare their invented situations. Question 5 is included to give students an opportunity to recognize a meaningless value for a probability.

Key Questions
Does it matter whether probabilities are written as fractions, decimals, or percents?

Why is the smallest probability 0?

Why is the largest probability 1?

What does a probability of 2.3 imply?

Pictures of Probability

Intent

In *Pictures of Probability*, students continue to develop tools for analyzing probabilistic situations. Students use area models, introduced using the notion of rugs, to determine theoretical probabilities. They also build on the skill of developing winning strategies. The counters game is based on rolls of two dice, and in the POW students develop a strategy for the nonprobabilistic game Linear Nim.

Mathematics

An **area model** represents the probability of an event as the ratio of two areas. If the region representing the event (the desired outcomes) is small compared to the region representing all possible outcomes (called the *sample space*), then the probability of the event is close to 0. If the desired region is almost the entire region, then the probability is close to 1. This tool allows students to find the theoretical probability of an event. For example, the following area model shows that the theoretical probability of getting two heads when flipping two coins is $\frac{1}{4}$.

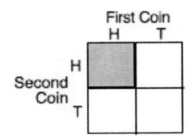

Progression

Pictures of Probability begins with a set of activities that introduce the area model for finding theoretical probabilities. Students then use this tool to answer questions about the results of rolling two dice. The second POW of the unit is also introduced.

Rug Games

Portraits of Probabilities

POW 4: Linear Nim

Mystery Rugs

The Counters Game

Rollin', Rollin', Rollin'

The Theory of Two-Dice Sums

Money, Money, Money

Two-Dice Sums and Products

Rug Games

Intent
Students are introduced to rug diagrams as a way to think about randomness and probabilities.

Mathematics
Students are introduced to an **area model** (using a rug metaphor) for determining theoretical probabilities. Area models employ the visual imagery of geometry to help students understand some abstract concepts, and, in particular, are useful for working with equally likely outcomes. The term *area model* will be introduced formally in the activity *The Theory of One-and-One*

Progression
Until now, students have worked with probability empirically or in the abstract. This activity introduces an area model as a concrete tool for thinking about theoretical probability. Initially, this model is presented using the metaphor of a rug.

Approximate Time
35 minutes

Classroom Organization
Groups, followed by whole-class discussion

Materials
Rug Games blackline master (handouts and transparency)

Doing the Activity
One way to introduce the activity is to tell this tale before students read the instructions.

I have a rug at my house, and there is a trap door in the ceiling directly over the rug. The trap door is the same shape and size as the rug. From time to time, the trap door opens and a dart drops directly down onto the rug. The process is quite random, which means that every point of the rug has as good a chance of getting hit as any other.

Now, of course, my guests never sit directly on the rug (it's dangerous!), but they like to sit nearby and guess which part of the rug the next random dart will hit. To keep things interesting, I have a variety of rugs of the same size that I can put out on different occasions.

Have groups look at the first rug in the activity. Ask, **Which color would you predict the dart is most likely to hit on this rug?**

As groups explain their predictions, you may want to have them show any additional lines they drew on the rug diagrams. For example, they might draw a diagram like this.

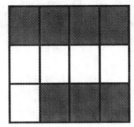

If students haven't yet used the language of probability to express their ideas, ask, **What is the probability of gray being hit? Of white being hit?**

Ask students to write the probabilities using the standard notation introduced in *What Are the Chances?* Thus they would write P(white) = $\frac{5}{12}$ and P(gray) = $\frac{7}{12}$.

As groups work through the rest of the activity, ask them to defend their predictions using area calculations and to use the standard notation for writing probabilities. Encourage the use of arguments based on the notion of *equally likely*. In the diagram above, a dart is equally likely to land in each of the 12 rectangles.

You might have the first groups to finish the activity create new rugs for each other.

Discussing and Debriefing the Activity

When students share their solutions to the rug problems, focus the conversation on understanding the presenter's reasoning rather than on whether the conclusion matches one's own.

Some students will probably break the rugs into equal-size pieces. You can relate this subdivision to the idea of equally likely outcomes. For instance, rug A can be divided as shown into 15 same-size pieces, each with an equal chance of being struck by a dart. Because 8 of the 15 pieces are gray, P(gray) = $\frac{8}{15}$. Similarly, P(white) = $\frac{7}{15}$.

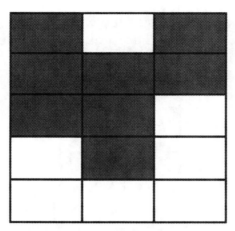

You can informally introduce the language of area into the discussion. For example, you can say that the gray region of rug A has a larger area than the white region.

You might remind students of the unit problem by mentioning that rug diagrams will eventually help them to evaluate strategies for the game of Pig.

Key Questions

Which color would you predict the dart to hit in this rug game?

What is the probability of the dart hitting gray? Hitting white?

Portraits of Probabilities

Intent
Students use area models to represent the outcomes of several probability situations. They also create situations to fit given area models.

Mathematics
This activity continues to reinforce the connection between area models (in the form of rug diagrams) and probabilistic situations.

Progression
Have students do this activity individually, likely for homework, following *Rug Games*. In class, students share ideas in order to build upon the variety of examples of probabilistic situations and associated area models.

Approximate Time
5 minutes for introduction

25 minutes for activity (at home or in class)

15 minutes for discussion

Classroom Organization
Individuals, followed by whole-class discussion

Doing the Activity
Part II of this activity is the first time students are given a rug diagram and are asked to describe a situation to match it. You may want to brainstorm a few possibilities for the rug diagram in Question 6. For example, the diagram might represent the probability that the tire that went flat on your new car was the front left tire.

Discussing and Debriefing the Activity

Part I: Rugs for Events
You might want to give transparencies and pens to five groups and have each group prepare a visual aid to lead the discussion about that situation. Each group may elect to synthesize its members' answers or use one particular member's answer.

As presentations are made, ask the rest of the class, **What is another way to draw a rug for this situation?**

For instance, Question 1 asks students to draw a rug with a shaded portion that represents a probability of $\frac{1}{6}$. Students will likely divide a rug into six equal sections, with one shaded. This could be done in various ways, such as those shown below.

Students could also draw a rug with just two sections, one representing $\frac{1}{6}$ and the other representing $\frac{5}{6}$, as illustrated below.

Students will probably explain Questions 2 and 3 using a theoretical model. Questions 4 and 5 bring up the extreme cases of a probability of 0 or 1.

Note that Question 3 is similar to Item D in *What Are the Chances?* in which students were asked to find the probability of flipping a coin twice and getting different results. Some students may still see the flipping of a coin twice as having three equally likely possible outcomes—two heads, one head and one tail, and two tails. It is important that they realize that these three outcomes are not equally likely.

In *What Are the Chances?* a list was suggested as a means of analyzing the two-coin situation. Now you might ask students how the two-coin situation can be represented with a rug diagram.

		Coin 1	
		H	T
Coin 2	H	Both heads	Coin 1: T Coin 2: H
	T	Coin 1: H Coin 2: T	Both tails

Drawing a rug diagram with four equal parts doesn't in itself explain why the four sequences are equally likely. The fact that the columns are of equal size represents an assumption that coin 1 is fair. The equal-size rows represent the fairness of coin 2. These fairness assumptions include the assumption that the tosses of coins 1 and 2 are **independent events**.

Displaying such a diagram may spark some insight and help to convince some students that the three outcomes—two heads, one head and one tail, and two tails—are not equally likely.

Using two different coins (such as a penny and a dime) can help students recognize that each rectangle represents a distinct outcome. Ask, **If you flip heads on the** dime, what might happen with the penny? How would you show that in a rug diagram?

Part II: Events for Rugs
As students share the situations they created, expect a variety of responses, including that some of the rug diagrams stumped some students.

Key Questions
What is another way to draw a rug for this situation?

If you flip heads on the dime, what might happen with the penny? How would you show that in a rug diagram?

Supplemental Activity
Heads or Tails? (extension) expands on the two-coin situation by asking students to investigate the probabilities for the outcomes of flipping three, four, and ten coins.

POW 4: Linear Nim

Intent
In this activity, students explore the idea of a winning strategy in another context.

Mathematics
The game of Linear Nim is distinct among other unit activities in that it does not involve chance. Variations of Nim have been traced far back in history. The game closely resembles the ancient Chinese game of Tsyanshidzi, or "picking stones," though its origin is uncertain. The earliest European references to Nim are from the beginning of the 16th century. Charles L. Bouton of Harvard University, who also developed a complete theory of the game in 1901, coined the name Nim. He never fully explained the origins of the name, which can be traced to German roots (nimm! meaning "take!") or the obsolete English verb *nim* of the same meaning. Notice that the word NIM upside down reads WIN.

Progression
Introduce students to the game approximately one week before their analyses and write-ups are due.

Approximate Time
15 minutes to introduce the game

1 to 3 hours for activity (at home)

20 minutes for presentations

Classroom Organization
Pairs to initiate the activity, concluding with presentations and class discussion

Doing the Activity
Allow some time for students to play the game in pairs.

During the week that students work on this activity, you may want to carve out a few minutes here and there for them to play the game again and to talk about their strategies. Ask, **When do you know you have won the game?**

The day prior to scheduled presentations, identify volunteers to present their findings. Give them transparencies and pens (or other resources) to prepare for their presentations. Some students may prefer to create a computer-based slideshow.

Discussing and Debriefing the Activity
Begin the discussion by having two or three students present their findings. During the follow-up discussion, focus on the theoretical analysis of specific strategies. It's

reasonable to expect most students to develop a complete strategy for the original game, and some may even develop a more general analysis.

One way to conclude the discussion is with a brief Nim tournament, randomly varying the initial and maximum values during the tournament. You will know that students really understand the game when they can determine as soon as the game is set up which player—the first or the second—will win. Their sense of accomplishment can be enhanced by using a random number generator to choose the parameters for each new game.

Key Questions

When do you know you have won the game?

Can you determine what you should do on your previous turn to make sure you can make that move next?

Supplemental Activities

Small Nim (reinforcement) This simpler problem may help students develop a strategy for winning Linear Nim. Consider assigning this activity a few days into students' investigation of this POW.

Piling Up Apples (extension) In this activity, students develop and evaluate strategies for winning a nonprobabilistic game similar to Linear Nim. Consider assigning this after the presentations on Linear Nim.

Mystery Rugs

Intent
In this activity, students continue to use rugs as a visual model for probability.

Mathematics
This activity reemphasizes that the distinct outcomes of a probabilistic situation must add to 1. It also incorporates a review of fractions, decimals, and percentages into further work with an area model.

Progression
Have students do this activity individually, likely for homework, following or together with *Portraits of Probabilities*.

Approximate Time
20 minutes for activity (at home or in class)

10 minutes for discussion

Classroom Organization
Individuals, followed by whole-class discussion

Doing the Activity
Little or no introductory discussion of the activity is necessary.

Discussing and Debriefing the Activity
Allow volunteers to share the situations they created for Questions 1 and 2, and ask the class to evaluate whether the situations fit the information. Help students find the missing probabilities ($\frac{7}{12}$ and .35, respectively) by asking, **What is the area of your other region? What does that tell you about the probability of the other outcome?**

Students may recognize that Question 3 presents an impossible situation, as the probabilities total to more than 1. If not, you can try to build on their work in finding the missing probabilities in Questions 1 and 2 to uncover the difficulty. If students find a solution to Question 3 by creating overlapping outcomes, commend them for their creativity.

Key Questions
What is the area of your other region? What does that tell you about the probability of the other outcome?

Can you name three (or more) outcomes for rolling a die whose probabilities add to more than 1?

The Counters Game

Intent
Students once again consider strategy while examining a game involving probabilities for the sum of two dice. This activity leads to the theoretical analysis of these probabilities.

Mathematics
The counters game provides another context for examining issues of strategy. The game is too complex for students to find the actual probability of success for each strategy. Through experimental analysis, however, students can build intuition about what strategies seem to work best. The context sets the stage for developing tools to determine theoretical probabilities for two-step situations.

Progression
Students first play this game in a group to get a sense of what might be a good or a poor strategy. Then groups come together to play a game as a class. Students will quickly recognize that some two-dice sums occur more frequently than others, motivating the need to derive the theoretical probability for each sum. This theoretical work will be done in *The Theory of Two-Dice Sums*.

Approximate Time
35 minutes

Classroom Organization
Small groups, followed by whole-class discussion

Materials
Dice (one pair per group)

Chips (11 for each student)

Sentence strips

Doing the Activity
Have students read the instructions for the game. Clarify that with each number rolled, every player removes one and only one counter from that numbered space. Each individual will need a strip of paper for a game board and 11 counters, and each group will need a pair of dice.

Before students begin, you might want to encourage them to first think of a strategy and to place their counters according to it. Allow time for groups to play several rounds and test various strategies. Suggest to them that learning the game and keeping accurate data to test how well each strategy works is more important that winning a particular round.

After groups have played a few games and students have written and thought about strategies, bring the class together for the competition proposed in Question 4. Each group will use a single game board and collectively decide on a strategy for placing counters. Have groups write down their strategies and place their counters accordingly. When groups are ready, begin the game by rolling a pair of dice.

Discussing and Debriefing the Activity

Ask each group to briefly explain its strategy for placing counters. Here are two additional questions you might ask:

Did some two-dice sums come up more often than others? Why?

Is it best to put all your counters on the two-dice sum that you think will come up most often?

Most students will see that some two-dice sums (such as 6, 7, and 8) come up more frequently and that others (especially 2, 3, 11, and 12) come up rarely. Their intuition will probably suggest that it makes sense to concentrate counters in the boxes with sums that come up more often.

A major goal of the activity is to get students interested in the following question, which you should raise if no one else does.

How might you find the theoretical probability of each two-dice sum?

Key Questions

What strategy did your group use?

Did some two-dice sums come up more often than others? Why?

Is it best to put all your counters on the two-dice sum that you think will come up most often?

How might you find the theoretical probability of each two-dice sum?

Supplemental Activity

Counters Revealed (extension) is intended for students who are interested in going deeper in their study of the counters game. The mathematical analysis of strategies for the game is quite complex and is not part of this unit. In this activity, students study a simplified version of the game. Students will do some experimental work in this activity, but will also be challenged to derive some reasoned explanations for their conclusions. Because of this activity's investigatory and open-ended nature, you may want to allow up to a week for work on it.

Rollin', Rollin', Rollin'

Intent

In this activity, students gather, graph, and summarize data concerning sums from rolling two dice. They continue to explore the usefulness of displaying data in frequency bar graphs. The collected data will confirm that some two-dice sums are more likely than others.

Mathematics

Students design a method to keep track of experimental results. Then, they devise a way to organize that information into graphical form.

Progression

As a follow-up to *The Counters Game*, students first discuss this activity as a class. They then collect data and complete the tasks individually, likely for homework. Once returning to class, they create a frequency bar graph of the class data and make observations. Finally, students calculate the empirically derived percentage for each two-dice sum for comparison with the theoretical analysis that will be done in *The Theory of Two-Dice Sums*.

Approximate Time

5 minutes for introduction

20 minutes for activity (at home or in class)

25 minutes for discussion

Classroom Organization

Individuals, followed by whole-class discussion

Materials

Dice

Graph chart paper

Doing the Activity

Stimulate interest in this activity by asking what techniques have been developed that may help students learn more about how frequently each two-dice sum occurs in the roll of a pair of dice. Some students will likely suggest actually doing the experiment and collecting data. Use this suggestion to engage students in developing a plan to collect data individually and then pool it together, as a large collection of data will provide a good sense of the theoretical probabilities.

Discussions about how to record and graph the data can occur before or after students begin their work. Involving students in the design of the data collection will help to alleviate the problem of some students not bringing legitimate data to class, if this activity is assigned as homework.

Discussing and Debriefing the Activity

You might begin the discussion by asking how students kept a record of their results. Some may have written down the results in the order they occurred, while others may have listed the possible outcomes and used tally marks to keep track. If no one mentions the second method, you might suggest it yourself.

Have volunteers share their graphs and any general conclusions they reached. If some students portrayed their information using something other than a frequency bar graph, ask them to share their representations as well. Presumably students' data contain only a few very high sums (10, 11, or 12) and a few very low sums (4, 3, or 2), with most of the sums in the middle (5 through 9).

The next stage is to gather all the data together to create a class frequency bar graph. You might ask, **What will we need to do to complete a frequency bar graph of all the data collected by our class?** To whatever degree is appropriate, engage students in designing and creating the class graph, using graph chart paper or a graph-paper transparency. Having each group compile its members' data and then send a representative to report to the class can be an efficient method. Students will need to think about the scale for the vertical axis, which will indicate how many times each outcome occurred. (You can expect about 250 7s in a class of 30 students.) The graph will be easier to read if column totals are written on the bars, as done below. This graph is from a simulation giving 50 results for each of 30 students, for a total of 1500 results.

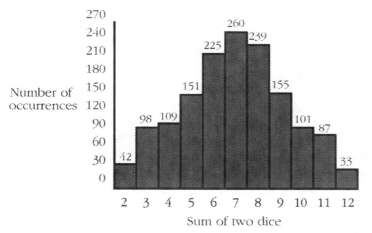

Generally, it is a good technique to begin conversations about data by asking, **So, what do you see?** Additionally, you might turn groups back for some discussion among themselves. After some group discussion, it is likely that students will see that the class data confirm many of their suspicions and may also reveal some things they didn't expect.

Use this opportunity to ask, **How does our class graph compare to your individual graphs?** This will probably elicit the observation that the class graph is "more even," "smoother," or "more symmetrical" than the individual graphs.

Finally, ask, **What do you think the sum of the column totals should be?** (It should be 50 times the number of students in class.) The class can verify that number and then use the total to find the percentage of occurrence for each sum. If the totals do not sum to a multiple of 50, some students will want to track down the error, a pursuit that may not be worth the necessary time.

Save the class frequency bar graph for comparison with the theoretical analysis to be done in *The Theory of Two-Dice Sums*.

Key Questions

What will we need to do to complete a frequency bar graph of all the data collected by our class?

How does our class graph compare to your individual graphs?

What should the sum of the column totals be?

The Theory of Two-Dice Sums

Intent
Students extend their use of area models to find the theoretical probability distribution for two-dice sums. In the process, they develop a tool for determining theoretical probabilities in a multistep situation.

Mathematics
The notion of a *probability distribution* is introduced through the use of frequency bar graphs. The term *rug diagram* is not standard among mathematicians. The transition to using the more common term **area model** can be gradual.

Progression
In this activity, students begin to design rug diagrams to help them determine theoretical probabilities. After some work in groups to develop a theoretical model, the class clarifies the model and makes connections between the theoretical probabilities and the data collected in *Rollin', Rollin', Rollin'*.

Over the course of the unit, students will work with probabilities associated with increasingly complex situations. Later they will encounter situations in which the outcomes are not equally likely and use two-dimensional diagrams to determine the probabilities.

Approximate Time
20 minutes for activity

30 minutes for discussion

Classroom Organization
Groups, followed by whole-class discussion

Materials
Dice in two colors (one pair per group)

Doing the Activity
Tell students that they will now develop a tool to determine the theoretical probability for each possible two-dice sum. You can start them thinking about the task by asking, **How would you draw a rug diagram to show what could happen after rolling one die?**

Give each group a red die and a white die (or two other colors) to begin working on the task. Using dice of two colors can help students understand, for example, that there are two ways to roll 3 and 5 (red 3 and white 5 or red 5 and white 3), while there is only one way to roll two 4s.

You will need to decide whether students are making progress on their own or whether it would be more productive to bring them together to share ideas. (You might choose to have students interrupt the work on this activity in order to complete the next activity *Money, Money, Money*, as that activity has only three possible outcomes, and then return to their work on this activity.)

Ask groups that are stuck what could happen after rolling 1 on the first die and how that could be shown in a rug diagram. **Can you think of a way to show the possible results on the second die, if you rolled 1 on the first die?** This question may encourage students to subdivide each of the six initial outcomes.

Discussing and Debriefing the Activity

Students may derive a diagram something like this.

		White die					
		1	2	3	4	5	6
	1	2	3	4	5	6	7
	2	3	4	5	6	7	8
	3	4	5	6	7	8	9
Red die	4	5	6	7	8	9	10
	5	6	7	8	9	10	11
	6	7	8	9	10	11	12

To emphasize the "equally likely" aspect of the problem, ask, **Which column is the most likely to occur? Which row? Which square is the most likely to occur when rolling two dice? Why are they all equally likely?**

Students should be able to articulate the idea that each square is equally likely, with a probability of $\frac{1}{36}$.

You can also ask for the probability of rolling individual combinations, such as 3 on the red die and 6 on the white die. Students should see that each combination has a probability of $\frac{1}{36}$.

You might also ask such questions as, **Which part of the diagram represents a roll of 4 on the white die?**

Next, ask students to identify the portion of the diagram that represents a roll totaling 6. The probability of getting 6 as the two-dice sum is $\frac{5}{36}$, because five of the 36 boxes contain a sum of 6.

By the end of the discussion, students should realize there are 36 equally likely outcomes for a pair of dice and be able to find the probability for each of the 11 possible sums (2 through 12).

Here are some additional questions you can pose to strengthen and clarify students' understanding of the area model:

Why are there two ways of getting a sum of 3, when the only way to get it is with a 1 and a 2? Students should see that "red = 1, white = 2" is a different square from "red = 2, white = 1."

What is P(even number)?

What is P(multiple of 5)?

Have students compute what the class totals would have been for *Rollin', Rollin', Rollin'* if the experiments had followed the theoretical probabilities. **What would the totals have been if our experiments had followed the theoretical probabilities perfectly?** Describe the answer as the *theoretical distribution* or theoretical probability distribution.

You might have students make a graph of this theoretical distribution and compare it to the graph of their experimental data. They will probably see some general similarity, but they will also see that their results vary somewhat from the theoretical distribution. You might ask, **Which is closer to the theoretical distribution—your individual data or the combined class data? Why?**

Post the rug diagram for two-dice sums for later use.

Key Questions

How would you draw a rug diagram to show what could happen after rolling one die?

Can you think of a way to show the possible results on the second die, if you rolled 1 on the first die?

Which column is the most likely to occur? Which row?

Which square is the most likely to occur when rolling two dice? Why (or why are all equally likely)?

What is the probability of getting a red 3 and a white 6?

Which part of the diagram represents a roll of 4 on the white die?

How does the diagram represent a sum of 6 on the two dice?

What is P(even number)? What is P(multiple of 5)?

What would the totals have been if our experiments had followed the theoretical probabilities perfectly?

Which was closer to the theoretical distribution—your individual data or the combined class data?

Supplemental Activity

Two-Spinner Sums (extension or reinforcement) is similar *The Theory of Two-Dice Sums*, but with a slightly different setting. In this set of two tasks, students examine the results of spinning two spinners and compare this work to the results of tossing two dice.

Money, Money, Money

Intent
Students examine another two-step probability problem. This activity can be done either during *The Theory of Two-Dice Sums* to help clarify the situation and draw out ideas, or after that activity to reinforce the techniques that emerged.

Mathematics
The two situations posed in this activity are two-step probability problems. If students are able to visualize the possible outcomes using an area model, it is likely they will be able to apply the techniques they have acquired to build an area model for *The Theory of Two-Dice Sums*.

Progression
Have students do this activity individually, likely as homework. During their work or the discussion, students should recognize that the two situations posed in this activity have a two-stage aspect to them that can be represented by the two-dimensional nature of a rug diagram (area model).

Approximate Time
20 minutes for activity (at home or in class)

20 minutes for discussion

Classroom Organization
Individuals, followed by whole-class discussion

Doing the Activity
This activity needs little or no introduction.

Discussing and Debriefing the Activity
For Question 1, students may see that it makes sense to set up the diagram with the result of one coin across the top and the result of the other coin down the side. With this diagram, they can show that the three outcomes Nina describes are not equally likely.

For Question 2, you can move students toward a diagram like the one below. This may not be the most obvious representation for some students, so expect the need for some clarification. You might ask, **What could happen if the $1 bill is pulled from the left pocket? How can the column showing the likelihood of drawing $1 be divided to show the three possible bills drawn from the right pocket?**

		Left pocket		
		$1 bill	**$5 bill**	**$10 bill**
Right pocket	**$1 bill**	$2	$6	$11
	$5 bill	$6	$10	$15
	$10 bill	$11	$15	$20

Students will probably see that each of the nine boxes is equally likely, and they can then find the probabilities.

Rather than explicitly discuss Question 3, you may just want to suggest that students incorporate the methods used in this activity into their two-dice sum analyses, looking for ways to represent equally likely outcomes.

Key Questions

What could happen if the $1 bill is pulled from the left pocket? How can the column showing the likelihood of drawing $1 be divided to show the three possible bills drawn from the right pocket?

From a probability perspective, is it the same or different to draw $5 from my right pocket followed by drawing $1 from my left pocket as it is to draw $5 from the right and $1 from the left?

Two-Dice Sums and Products

Intent
In this activity, students determine some simple two-step probabilities and create related probability questions of their own. It is an opportunity to practice the mathematics developed in recent activities.

Mathematics
This activity concludes a sequence designed to solidify the basic definition of probability and introduce area models as a tool for understanding complex probability problems.

Progression
Have students do this activity individually, likely as homework. Small-group and class discussions should provide an opportunity to share answers, methods, and reasoning and confirm student confidence in being able to solve similar problems using an area model.

Approximate Time
30 minutes for activity (at home or in class)

15 minutes to share ideas

Classroom Organization
Individuals, followed by small-group and whole-class discussions

Doing the Activity
This activity will need little or no introduction.

Discussing and Debriefing the Activity
Students may have found the answer to Question 1 by adding up the number of cases giving each possible even result (or each odd result). They would find one way to get 2, three ways to get 4, and so on, for a total of 18 ways to get an even result. The probability of an even sum is thus $\frac{18}{36}$, which simplifies to $\frac{1}{2}$.

Similarly, students might add the appropriate probabilities, in this case P(2) + P(4) + P(6) + P(8) + P(10) + P(12). So, the probability of an even sum is $\frac{1}{36} + \frac{3}{36} + \frac{5}{36} + \frac{5}{36} + \frac{3}{36} + \frac{1}{36} = \frac{18}{36} = \frac{1}{2}$.

This is a good opportunity to point out that if students get a simple answer such as $\frac{1}{2}$ after performing some complicated arithmetic, they might look for an easier way to find the answer. In this case, they might notice that half the sums in each row of the rug diagram are even.

For Question 3, students will have to go through an analysis similar to that done for *The Theory of Two-Dice Sums*. There are more possibilities for two-dice products than for two-dice sums. The most common products are 6 and 12, each with a probability of $\frac{1}{9}$.

For Question 4, students should see that even products are more common. You may want to ask, **Why are odds so much less common as products than as sums?** Someone may be able to articulate that a product is odd only when both factors are odd, which happens only one-fourth of the time when rolling dice.

Key Question
Why are odds so much less common as products than as sums?

Supplemental Activities
Different Dice (reinforcement) is similar to *Two-Dice Sums and Products* and brings out the notion that the sum of all the probabilities in a given situation must equal 1. You might use it with students who need more practice with ideas like those in that assignment.

Three-Dice Sums (extension or reinforcement) explores the same ideas in a more complex case.

In the Long Run

Intent

Expected value—the average value over the long run—is the focus of the activities in *In the Long Run*. Students develop ways of computing expected value in a variety of situations. They continue to use area models, and are introduced to tree diagrams, in their analyses of the long-run behavior of probabilistic events.

Mathematics

For the rest of this unit, the notion of *in the long run* will be emphasized to ground students' reasoning about the concept and the computation of **expected value**, essential to analyzing strategies.

Rolling a die is a random process; no one can predict the outcome of any single roll. However, in the long run we can predict with confidence that each number will come up on $\frac{1}{6}$ of the rolls. The sum of the numbers is 21, so in the long run the average number per roll is $\frac{21}{6} = 3.5$. *Expected value* is shorthand for *average in the long run*. The expected value per roll of a die is 3.5.

Expected value ties together theoretical and observed probabilities. The expected value is a theoretical result; in the case of a die, no roll can result in 3.5. But expected value is the best prediction of an observed probability.

One source of observed probabilities is simulations of random processes. Students will use spinners, dice, and playing cards to conduct several simulations in *In the Long Run*. They also continue to use area models for analyzing multistage situations, and they are introduced to tree diagrams.

Progression

Over the course of *In the Long Run,* students will examine several games in which the payouts to the players are determined by the results of some random event. Methods for quantifying results are made more formal in these activities, in anticipation of returning to the game of Pig. The final POW of the unit, in which the focus is once again on strategy analysis, is posed early in *In the Long Run*.

Spinner Give and Take

Pointed Rugs

POW 5: What's on Back?

Mia's Cards

A Fair Rug Game?

One-and-One

A Sixty Percent Solution

The Theory of One-and-One

Streak-Shooting Shelly

Spins and Draws

Aunt Zena at the Fair

Simulating Zena

The Lottery and Insurance—Why Play?

Martian Basketball

The Carrier's Payment-Plan Quandary

A Fair Deal for the Carrier?

Simulating the Carrier

Another Carrier Dilemma

Spinner Give and Take

Intent
In this exploratory activity, a spinner game is used as a probability model to introduce the notion of **expected value**. The approach is to consider a large number of trials in order to determine the average *over the long run*.

Mathematics
This activity draws upon students' intuitive understanding of the **law of large numbers** as they consider what will happen in a game involving a back-and-forth payoff between two players that is determined by set probabilities. Students are asked to consider who will come out ahead over the long run. This experiential task begins their investigation into expected value, a description of the average points won (or lost) per turn over the long run. In this unit, the notion of *over the long run* will be emphasized to ground students' reasoning about the concept and the computation of expected value.

Progression
Students begin by predicting the winner of a game and then carry out simulations to test their conjectures. Through this experience, students begin to develop a theoretical analysis of the game's structure. Midway through the investigation, the class comes together to share ideas and to consider some of the quantitative conclusions that can be drawn.

Approximate Time
35 minutes

Classroom Organization
Pairs, followed by whole-class discussion

Doing the Activity
This activity is best completed in pairs. Demonstrate how to make a spinner, if necessary.

Discussing and Debriefing the Activity
Students will often neglect to consider that Al and Betty are paying each other in the game. This observation can be used to reengage students who are quick to decide they have finished the activity. Ask, **If Betty gets her winnings from Al, what effect does that have on Al's bank account (or wallet)?**

Begin the discussion when all pairs have finished Questions 1 and 2.

For Question 1, ask several pairs to share their predictions and reasoning. How did you make your predictions for Question 1? Roughly speaking, students should articulate that although Betty wins more often, Al wins more money each time he wins.

Students should be expected to do some sort of quantitative work to support their predictions. They may do this in terms of a specific number of spins (for example, 100), estimating how many times each player would win and how much money each ends the game with.

For Question 2, ask several pairs for their results. Some pairs may have ended with Betty ahead, though most will have Al ahead. Ask if the fact that some pairs have Betty winning after 25 spins means their reasoning in Question 1 was incorrect. If necessary, call students' attention to the phrase *in the long run* in Question 1. Students should gradually be developing and articulating the idea that although any result can occur in a small number of spins, *in the long run* the number of times each person wins will be roughly proportional to the probabilities.

Consider pooling all the pairs' results; presumably Al will win about $\frac{1}{4}$ of the time and Betty about $\frac{3}{4}$ of the time. You might return to these numbers after the discussion of Question 3 to compute each player's total winnings.

Answering Question 3 involves moving from probabilities to number of wins to amounts won to the difference in those amounts. Follow students' lead in the discussion while you guide them toward an analysis that goes something like this.

1. What is Betty's probability of winning? Al's? From the shading of the spinner, Betty's probability of winning a given spin is $\frac{3}{4}$ and Al's is $\frac{1}{4}$.

2. How many times out of 100 should Betty win? What about Al? If Betty and Al spin 100 times altogether, and the results follow the theoretical probabilities, we can expect Betty to win about 75 times and Al to win about 25 times.

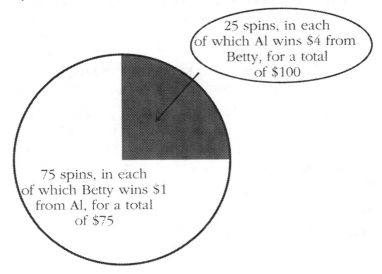

3. How much do they each win? Betty would win $1 from Al in each of her 75 victories, and Al would win $4 from Betty in each of his 25 victories. So, in 100 spins, Betty will win about $75 from Al (75 • $1) and Al will win about $100 from Betty (25 • $4).

4. How far ahead will Al be at the end? Over the course of 100 spins, Al will be about $25 ahead of where he began.

In the discussion, emphasize that in any particular set of spins, a variety of results are possible. You might ask, How many spins would Al have to win in order to come out ahead? Students can use guess-and-check to see that he needs to win at least 21 out of 100 spins to come out ahead of Betty.

The fact that Al wins even if the results are slightly off from the probabilities is part of the strength of the "large number of spins" approach. Students should gradually develop an intuitive sense that the greater the number of spins, the more likely it is that Al will come out ahead.

Have students conduct the same analysis with at least two different numbers of spins. They should see in each case that Al comes out ahead and that he wins more money if they play more spins. In fact, for twice as many spins, he wins twice as much.

You can ask what remains the same in each case, besides the fact that Al is ahead. You might ask them to consider what it may mean that, for example, for twice as many spins, this analysis shows Al winning twice as much. This question addresses the idea directly: How much did Al win per spin, on the average? Students should see that in each case, the average is 25¢. In other words, his end result is the same as if he had won 25¢ on each spin.

In the upcoming activity *Mia's Cards*, the term expected value is introduced as a shorthand for the idea of *average per spin* (in the long run). For now, the goal is that students begin to agree that *average per spin* is a good way to measure what happens in the long run, as the outcome is the same no matter how many spins are made.

Ask, **How does this problem relate to the game of Pig?** You may want to use this question as a topic for focused free writing and then have students share their ideas. As needed, remind students that in the game of Pig, they are also interested in what happens in the long run. Specifically, they want to know which strategy is likely to produce the most points per turn in the long run. You can mention that finding the probabilities for a Pig strategy is much harder than finding the probabilities for this simple spinner game.

Key Questions

If Betty gets her winnings from Al, what effect does that have on Al's bank account (or wallet)?

How did you make your predictions for Question 1?

What results did you get for Question 2?

IMP Year 1, The Game of Pig Unit, Teacher Guide
© 2009 Interactive Mathematics Program

Does the fact that some pairs showed Betty winning mean that the reasoning was incorrect?

What is Betty's probability of winning? Al's?

How many times out of 100 should Betty win? What about Al?

How much do Betty and Al each win?

How far ahead is Al at the end?

How many spins would Al have to win in order to come out ahead?

What happens if you use a different total number of spins?

How much did Al win per spin, on the average?

How does this situation relate to the game of Pig?

Pointed Rugs

Intent
In this activity, students continue to apply the technique of considering a large number of trials to analyze probabilistic situations involving points or payoffs.

Mathematics
Students build on the notion of considering a large number of trials to analyze situations with weighted probabilities. They continue to use an area model to think about probability and consider why it is possible for an outcome with a lower probability to result in more points over the long run.

Progression
After working individually, students share methods and demonstrate their reasoning. The method of comparing the total points accumulated after some large number of trials is emphasized. Students see that different "large numbers of games" yield the same winner and that a convenient number can be selected for this analysis. These strategies are further developed and drawn upon through the remainder of the unit.

Approximate Time
25 minutes for activity (at home or in class)

15 minutes for discussion

Classroom Organization
Individuals, followed by whole-class discussion

Materials
Rug Games blackline master (transparency, optional)

Doing the Activity
Explain that this activity uses the rugs from *Rug Games*, so students can use the probabilities found for that activity. However, the situation is now more complex, because each region now has points associated with it.

Discussing and Debriefing the Activity
You might assign one or two groups to each rug and have them prepare a presentation of their solution to initiate class discussion. During this time, observe groups sharing methods, and note who has done the homework and how well students seem to have understood the ideas in the assignment.

If groups are having difficulty, you might suggest they pick a convenient number of games to play and see which color would earn the most points if the darts landed according to probability. This "large number of trials" method is used throughout the unit, so be sure it is among the methods presented.

Some students may break up the area into equal-size sections, write the appropriate number of points in each section, add up the total points, and divide by the number of sections (as if each equal section got one dart).

Following up on the "large number of trials" method, ask, **Did you all use the same number of games?** Since the answer will likely be no, you can bring out that the predicted winning color is the same no matter how many games are played, even though the total number of points earned for each color will vary with the number of games.

As students begin to value this approach, help them to see that there is often a convenient "large number of games" to select. Ask, **Do some numbers of games work better than others?** As needed, you can ask about the advantages of choosing certain numbers in a particular situation. For example, for rug A the probabilities are $\frac{7}{15}$ and $\frac{8}{15}$, so it would be convenient to choose a large number that is a multiple of 15.

As with the discussion of *Spinner Give and Take,* you may want to bring out that although the results usually won't follow the probabilities exactly, the fraction of darts landing in each color should be close to the theoretical probabilities in the long run. If the total number of darts is large, slight variations from the theoretical probabilities won't change which color yields the most points. A numerical example might help students grasp this idea.

Key Questions

Did you all use the same number of games?

Do some numbers of games work better than others?

POW 5: What's on Back?

Intent
In this POW, students use both experimental and theoretical methods to examine a probability situation. Their success in this activity should be measured in terms of the growth in their understanding of the concepts involved.

Mathematics
This activity provides a context for examining issues of strategy. Although no strategy will always ensure a win in this game, students can determine the strategy that optimizes their chance of winning. Students carry out trials to approximate the experimental probability for each of several strategies and apply techniques developed in the unit to determine a theoretical probability for winning. Students use their intuitive notions about the law of large numbers to justify the reasonableness of relying on experimental probabilities and to confirm the accuracy of the theoretical probabilities.

Progression
Students begin by considering a seemingly simple game in which a player is to guess "what's on back" of a card pulled from a bag. Asked to determine the strategy with the highest probability of success, students must design experimental and theoretical analyses to test various strategies. The problem is counterintuitive for most people and can be difficult to reconcile in a short period of time. As always, students should work over the course of the week and come together midway to share ideas and discoveries. Students will present their analyses for two given strategies and any others they explore.

You may seek to conclude the discussion by generalizing the possible strategies in order to determine the best strategy. Similar reasoning will be used later in the unit as students return to attempting to find the best strategy for the game of Pig.

Approximate Time
10 minutes for introduction
1 to 3 hours for activity (at home)
10 minutes for follow-up midweek
20 minutes for presentations

Classroom Organization
Individuals, followed by small-group discussions, concluding with whole-class presentations

Materials
Bags
Heavy cardstock

Doing the Activity

This POW involves complex ideas about strategy and probability. You might model the game and give students some time to play it. Explain that everyone will need to make a set of cards. Bring out these ideas during the introductory discussion:

A strategy must tell what to do in response to what is shown on the drawn card.

Rather than just looking for the *best* strategy, students are to analyze *several* strategies.

Each experiment involves drawing a card, making a prediction based on the strategy being tested, and recording the accuracy of that prediction. To help students understand the value of this methodical analysis, you might ask, **In this POW, what does it mean to experiment?**

Doing just a few experiments is not very meaningful. In *Expecting the Unexpected*, students saw that even 50 flips didn't guarantee an accurate reflection of the theoretical probabilities. Bring up this idea by asking, **How many experiments might you need to do?** Since students have no theoretical basis for judging how many experiments might be needed, talk about how they can determine how many are enough. You may want to call their attention to the guideline on sample size in the POW itself, which suggests that they continue until their "overall results begin to stay roughly the same" as they repeat the experiment. (Students won't fully understand the significance of sample size until later in their studies, but they should begin to appreciate it intuitively.)

You may also want to talk about the difference between experimental results and results derived from a theoretical model. Although an experiment gives a feel for what the results might be, even a large number of trials does not guarantee an accurate picture of the theoretical probability. If the theoretical analysis gives a result very different from the experimental estimate, this might indicate an error in the theoretical analysis. The following prompt may help initiate this discussion: **What are the advantages of experiments in finding probabilities? What are the advantages of theory?**

As always, monitor and encourage student work throughout the week. If students are having difficulty starting, you might focus them on the first strategy described in the POW (predicting that the mark on the other side will be different from the mark showing).

It's good if students realize that they need to do more than a handful of experiments to get an accurate estimate of the probability of success for a particular strategy. Often students feel as though they are "doing the problem" after completing only a few experiments. Conversation and comparison of experimental data may be necessary for them to gain a greater appreciation of the need to do more trials. Ask, **How many experiments are you finding you need to do?** to encourage reflection on the notion of continuing until the experimental probabilities stabilize.

If needed, remind students that they are also looking for a *theoretical* analysis of the probability. Ask, **How might you develop a theoretical model?** In this unit, they have seen theoretical probabilities derived through rug models.

It is a common misconception that if one initially sees **X,** the probability is .5 that the other side is also **X.** This idea comes from the fact that there are only two cards containing an **X**, which suggests that the chosen card is equally likely to be one of these two cards. One of these cards has **X** on the other side, and one has **O** on the other side. However, the truth is that if you're looking at a side containing **X**, you're more likely to have drawn the card with an **X** on both sides than the card with an **X** on only one side. You might wish to draw attention to this concept after students have worked for a short time. The discussion can highlight the value of using experimental probabilities to test the reasonableness of proposed theoretical probabilities.

As the day for presentations draws near, you might ask for volunteers to present, with each focusing on a different aspect of the problem. For example, one might talk about results with one strategy, another about results with a second strategy, and a third about finding the best strategy.

Discussing and Debriefing the Activity

This POW may have been difficult for many students. Although the content of this activity is related to that of the unit, students certainly don't need to uncover the best strategy or master the mathematical content to be successful with the activity or the unit. The intent of POWs is to develop problem-solving skills and positive attitudes. Even if students come away from this problem without understanding why a particular strategy is best, they will probably have strengthened their appreciation of several ideas, including

- using an experiment to estimate a probability
- understanding that the accuracy of an experimental estimate will generally depend on how many times the experiment is done
- recognizing that area models are a useful tool for finding theoretical probabilities

During the presentations, focus the discussion on students' work with the two strategies suggested in the activity. Bring out the idea that an experiment will give a more accurate estimate if it is repeated more often.

In discussing how to determine the best strategy, you may first want to identify what the possible strategies are. Focus students' attention on strategies that are based on what shows up on the side of the card drawn from the bag. From this perspective, each strategy should look something like:

If you get a card with **X** showing, then predict ____.

If you get a card with **O** showing, then predict ____.

As presenters discuss their theoretical analyses, keep in mind that a crucial element of the theoretical analysis of any strategy is determining what the equally likely outcomes are. One approach is to recognize that each card is equally likely to be drawn and, then, that each side of the drawn card is equally likely. This is equivalent to saying that each of the six sides is equally likely to be the initial side viewed. A diagram like the one below can then be very useful in examining strategies. Here, sections above or below each other represent the two sides of a given card.

X	X	O
X	X	O

This diagram can be shaded to indicate cases in which the first strategy, "predict different," would be successful. The top left box below is unshaded because if you draw the card and observe the face it represents, the other side will be the "same." The bottom middle box is shaded because if you select that card and observe the **O** face, the other side will be "different." Finish the diagram by considering the four remaining boxes.

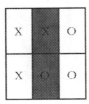

So, for the "predict different" strategy, P(success) = $\frac{1}{3}$.

A similar analysis for the strategy "always predict **X**" produces the diagram below and gives P(success) = $\frac{1}{2}$. The strategy "always predict **O**" also gives P(success) = $\frac{1}{2}$, so either of these strategies is better than the first one.

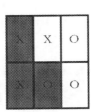

Finally, the strategy of "predict same" is represented in the next diagram. For this strategy, P(success) = $\frac{2}{3}$.

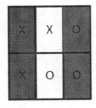

Thus, "predict same" is the best possible strategy.

Key Questions

In this POW, what does it mean to experiment?

How many experiments might you need to do?

How many experiments are you finding you need to do?

Supplemental Activity

Make a Game (extension) asks students to work in pairs to create games that involve probability and strategy. You may want to provide a clear schedule for the various stages of this activity, such as choosing partners and presenting preliminary plans. Make the grading criteria clear (such as use of probability and the need for strategy, entertainment, and clear instructions), perhaps even giving students a voice in establishing those criteria. You may want to provide class time for students to work on their games and allot a full day toward the end of the unit for them to share and explain their games.

Mia's Cards

Intent
Students analyze another probabilistic situation involving a payoff by considering a large number of trials. In the discussion of the activity, they are introduced to the term *expected value*.

Mathematics
The "large number of trials" approach used in *Spinner Give and Take* and *Mia's Cards* gives the concept of **expected value** a solid intuitive foundation. The term *expected value* is introduced in the discussion of *Mia's Cards*, with the acknowledgment that this is simply a formal name for an idea that students have already encountered in the context of examining averages *in the long run*.

Over the course of the activities in *In the Long Run*, students apply the concept of expected value in increasingly complex situations. Near the end of the unit, they will use expected value to evaluate strategies for a game similar to Pig, in preparation for justifying the best strategy for Pig.

Progression
Students work individually on the activity and then come together to share methods and ideas. The teacher then introduces the term *expected value* for the concept of computing an average per turn over the long run. Students will connect the concept to previous work in the unit and to the unit problem: finding the strategy for the game of Pig that gives the highest expected value.

Approximate Time
5 minutes for introduction

20 minutes for activity (at home or in class)

30 minutes for discussion

Classroom Organization
Individuals, then groups, followed by whole-class discussion

Doing the Activity
This activity is similar to the earlier spinner and rug games, though students may find it more difficult, as the situation is more complex. If assigning the activity as homework, make sure students know the number of cards and suits in a typical deck.

Discussing and Debriefing the Activity
Have students share their results in groups. As they are doing so, you might give some groups transparencies, assign one or two groups each to Questions 1 and 2,

and have other groups choose a game that one of their members invented for Question 3.

Presentations of solutions to Questions 1 and 2 can help students feel comfortable working with a large number of games, finding the total number of points, and then finding the average by dividing by the number of games.

Presentations that use different numbers of games will help dramatize the point that the average will be the same. For example, in Question 1, using 100 games gives Mia about 25 hearts, for 250 points, and about 75 cards from the other suits, for 375 points. Using 500 games gives her about 1250 points from hearts and 1875 points from the other suits, for a total of 3127 points. Each total gives an average of about 6.25 points for each card picked.

Introduce the term expected value for "the average amount gained (or lost) per turn in the long run." Ask students to restate their results from Questions 1 and 2 using this term. For instance, for Question 1, students might say that Mia has an expected value per turn of 6.25 points. (For emphasis, it's a good idea to use a phrase like *per game* or *per turn* whenever you use the term *expected value*.)

Help students see that expected value is an average in the long run, not what one expects on any particular trial. Ask, **So Mia will get 6.25 points each time she draws a card?** You may want to emphasize that expected value is not a new concept, but simply a concise name for an idea students have already worked with. You can point out that the use of such terminology makes it easier to state complex ideas.

Students can gain practice with the terminology by rephrasing their results from *Waiting for a Double* and from Question 3 of this activity. Ask, **Can you rephrase other problems you've seen in terms of expected value?** Students should notice that their work in *Waiting for a Double* gave an experimental estimate of the expected value for the number of rolls needed to get a double.

If students are having difficulty, use some of the games they made up for Question 3 to solidify the idea. Even if they seem comfortable with the concept of expected value, you might ask groups to share some of their games and have the class figure the expected value of each game.

If students haven't already made the connection between the concept of expected value and the game of Pig, you might ask, **How can you restate the unit problem in terms of expected value?** Help students to realize that they are looking for the Pig strategy that gives the highest expected value. Edit or add to the posted unit goal to reflect this use of the term.

Key Questions

Does an expected value of 6.25 points per turn mean that Mia will get 6.25 points each time she draws a card?

Can you rephrase other problems you have seen in terms of expected value?

How can you restate the unit problem in terms of expected value?

Supplemental Activities

Pointed Rug Expectations (reinforcement) offers more experience with the concept of expected value, using the diagrams from *Pointed Rugs*.

Explaining Doubles (extension) asks students to return to *Waiting for a Double* and explain the expected values found in that activity.

Two Strange Dice (reinforcement or extension) uses a pair of nonstandard dice and requires students to consider various combinations of outcomes in computing an expected value. It also suggests the idea of multiplying each possible numerical outcome by its probability and adding the results to get the expected value.

Expected Conjectures (extension) asks students to examine, in an intuitive way, the principle that the "large number of trials" method for computing expected value is actually independent of the number of trials.

Squaring the Die (extension) explores how expected value is affected when events are combined.

A Fair Rug Game?

Intent
This activity introduces the notion of a **fair game** and the use of negative numbers for expected value.

Mathematics
In this activity, students consider negative values for expected value and what it means for expected value to be zero. Students will use expected value to compare strategies later in the unit, considering which strategy will yield the most points per turn on average over the long run.

Progression
After students have completed their work individually, the discussion should first identify each player's expected value and the fact that one expected value is the opposite of the other. Next, a discussion of a **fair game** will set the stage for altering the scoring to make the game fair. Identify that when the game is fair, each player's expected value is now zero.

This activity continues to build students' fluidity with deriving expected value and understanding its meaning. Students should continue to emphasize a "large number of trials" approach.

Approximate Time
5 minutes for introduction

20 minutes for activity (at home or in class)

20 minutes for discussion

Classroom Organization
Individuals, followed by whole-class discussion

Doing the Activity
This is the first time students will consider a fair game. You may want to explain that in a fair game, both players are expected to come out equally well in the long run.

Discussing and Debriefing the Activity
Have a volunteer begin the discussion of Question 1. As needed, help students move beyond the fact that Tony wins $\frac{9}{15}$ of the time to focus on how *much* the players win. Students will likely use a "large number of trials" analysis to find that Tony loses, say, $30 over 150 games.

As the conversation focuses on Tony's expected value per turn, the idea of expected value being a negative number will probably occur to many students. Confirm that this is the standard way to express the idea that, over the long run, Tony loses money. Tony's expected value per turn can be written as –$0.20 or –20¢.

Through their calculations, students will see that Crystal's expected value is the opposite of Tony's.

Once the two expected values have been determined, focus on the question of whether the game is fair. Ask, **What does "fair" mean in the context of this game?** Students will probably agree that a fair game is one in which the players come out even in the long run. This game is not fair, because Crystal comes out better in the long run.

If the game were fair, what would Tony's expected value be? Some students may be ready to answer this question. If not, you might then ask, **How could you change the game to make it fair?** Let volunteers offer suggestions. The smallest whole-number solution for the payoffs is for Tony to win $2 from Crystal when the dart lands on black and Crystal to win $3 from Tony when the dart lands on white. Any payoff in this 2:3 ratio will work.

Have students now confirm that if the game were fair, Tony's and Crystal's expected values would both be $0.

Key Questions

What does "fair" mean in the context of this game?

If the game were fair, what would Tony's expected value be?

How could you change the game to make it fair?

Supplemental Activity

Fair Spinners (reinforcement) asks students to modify the spinner from *Spinner Give and Take* to create another fair game.

One-and-One

Intent
This activity sets the stage for a several-day investigation of a more complex probabilistic situation. Students apply the concept of expected value to this new situation, this time in relation to *most likely outcome*. They conduct simulations as well.

Mathematics
Over the course of the activities in *In the Long Run,* students apply the concept of expected value in increasingly complex situations. *One-and-One* presents a two-stage process in which the outcome of the first stage determines whether there is a second stage. *Streak-Shooting Shelly* also involves two stages, with the probabilities changing in the second stage. The situation in *Martian Basketball* has up to three stages.

One-and-One employs the first focused use of a simulation, in which students model a one-and-one situation in basketball by pulling colored cubes from a bag.

Progression
This activity introduces the one-and-one free throw context, which is used in the next several activities. Students first guess how many points they think are most likely for a one-and-one free throw shooter to make and then design a simulation to test their guesses. In the next activity, *The Theory of One-and-One,* students develop theoretical methods to analyze their initial guesses. Other activities are intermingled for student practice and extension.

Approximate Time
25 minutes

Classroom Organization
Whole-class discussion, with some data generated in pairs

Materials
Large collection of red and yellow cubes (or similar items in two colors)

Small paper bags

Doing the Activity
Ask whether anyone can explain what a one-and-one situation is in basketball. A one-and-one occurs in a penalty situation in which one player has committed a foul against another. The player who has been fouled is allowed to take a free throw. If the free throw is unsuccessful, the one-and-one situation is over. If the first shot is successful, the player gets to shoot once more. Each successful shot scores 1 point.

The player can thus score a total of 0 points (by missing the first shot), 1 point (by making the first shot but missing the second), or 2 points (by making both shots).

Once students understand the one-and-one situation, have them read the problem described in *One-and-One*, in which Terry has a 60% chance of success on each attempted free throw. Clarify the meaning of this by asking how many shots Terry would be likely to make out of 100, out of 40, out of 15, and so on.

Then have students, working individually, consider the question posed in the activity: **In a one-and-one situation, how many points is Terry most likely to score: 0, 1, or 2?** When they have thought about the question, conduct a class vote, recording the number of students who select each possible score.

Tell students that they will now design and conduct an experiment to estimate the probabilities involved in the situation. Introduce the word **simulation**. Ask, **Where have you heard the word simulation before?** (One example is flight simulators.) **Why do people use simulations?** The main idea that should emerge is that a simulation allows us to learn about something when we can't investigate that thing directly. You might also ask, **Have you done any simulations for other mathematics problems in this unit?** Students might mention their current work on *POW: What's on Back?*

Ask students to describe the difference between finding probability using experimental results and using a theoretical analysis. They should be growing more comfortable articulating that an experiment can give them a feel for what the results might be, while a simulation will give only observed probabilities, which can at best approximate theoretical probabilities, even with a large number of trials.

Give students the paper bags and cubes and ask, **How might you use these materials to set up a simulation to study the question in this activity?** Alternatively, you might ask students to design a simulation without prompting them with specific materials. They might suggest using other objects or propose using a random number generator.

If students suggest using 60 cubes of one color and 40 of the other, ask whether there is a smaller number of cubes that would work. Some students will identify 3 and 2; others will be more comfortable with 6 and 4. Focus on why the particular combinations they identify are appropriate. Ask, **Why is this combination of cubes suited to this problem?** Students should be able to articulate that if 60% of the cubes are red, picking a red cube represents making a free throw and picking a yellow cube represents missing a free throw.

It's a good idea to have students try to describe exactly how they will conduct the simulation. For example, they might construct instructions like these:

Shake the bag and pull out a cube. If the cube is yellow, the simulation is over. Write 0 for the score. If the cube is red, draw again to complete the simulation. (Return the red cube to the bag and shake before drawing again.) If the second cube is yellow, write 1 for the score. If the second cube is red, write 2 for the score.

Conduct a few simulations as a class, and then have students work in pairs. Each pair should simulate the one-and-one situation approximately 20 times, recording each result, to gather lots of data for the class to examine. When they are finished, have each pair tally the number of 0s, 1s, and 2s they got.

Discussing and Debriefing the Activity

Compile a class total of the 0s, 1s, and 2s from the simulations. As a class, calculate the percentage of the time each score occurred. If the theoretical probabilities are borne out, the experiment will have produced more 0s than any other number. (However, 2s are a close theoretical second.)

Typically, most students will have guessed that a score of 1 is the most likely. This conversation will be continued in the discussion of *Theory of One-and-One,* when students observe that the expected value per one-and-one situation is very close to 1, even though the score of 1 is actually the least likely result.

Ask, **Do you think the simulation is a good method for analyzing the situation?** Tell students that they will later use an area model as a theoretical model to analyze the situation.

Key Questions

In a one-and-one situation, how many points is Terry most likely to score: 0, 1, or 2?

Where have you heard the word simulation before? Why do people use simulations?

Have you done any simulations for other mathematics problems in this unit?

What is the difference between experimental results and theoretical analysis?

How would you set up a simulation using these materials?

Why is this particular combination of cubes suited to this problem?

Do you think the simulation is a good method for analyzing the situation?

A Sixty Percent Solution

Intent
In this activity, students design their own simulations and are encouraged to think more deeply about the difference between "average over the long run" and "most likely to occur."

Mathematics
The discussion of students' simulations may bring forth the notion that as more experiments are conducted, the outcomes tend toward the theoretical probabilities. The activity also indirectly inquires about the expected value—preparing students to wrestle with the difference between "average over the long run" and "most likely to occur."

Progression
This activity should be done individually, followed by a brief whole-class sharing and discussion of results. The activity is very similar to students' work in *One-and-One* and sets the stage for developing methods to figure two-stage probabilities in *The Theory of One-and-One*.

Approximate Time
5 minutes for introduction

20 minutes for activity (at home)

10 minutes for discussion

Classroom Organization
Individuals, followed by whole-class discussion

Doing the Activity
Tell students that they will be designing a simulation at home. You might encourage some brainstorming of materials they might use.

Discussing and Debriefing the Activity
Ask a few volunteers to report their results. Then ask, **Was the average score you found also the most likely result?** Although the majority of students should have found that Terry's most common result was 0 or 2, most should also see that the average for 40 trials is quite close to 1.

Ask students how they could state the results for Question 4 in terms of expected value. They should see that the answer to Question 4 is an experimental estimate of Terry's expected value for each one-and-one situation.

Key Questions

What results did you get?

Was the average score you found also the most likely result?

Can you state the results for Question 4 in terms of expected value?

The Theory of One-and-One

Intent
In this activity, students complete a theoretical analysis of the one-and-one situation. In the process, they expand their use of area models to determine the theoretical probabilities in a multistage event and to compute expected value.

Mathematics
The theoretical analysis of the one-and-one situation continues to develop students' understanding of expected value—not only its computation, but also its meaning as an average over the long run, as opposed to what is most likely to occur. Students also continue their use of the area model as a tool to figure the probabilities of a multistage event.

This understanding of probability and expected value will be used in the final analyses of the games of Little Pig and Pig itself.

Progression
Students will begin this activity in groups. When some groups have completed the activity, the class will come together to share ideas about techniques and the answers they yield.

Approximate Time
40 minutes

Classroom Organization
Small groups, followed by whole-class discussion

Doing the Activity
Students have completed an experimental analysis of the one-and-one situation. Now they will work in groups to develop a rug diagram analysis of Terry's expected value for each one-on-one situation. (You may want to begin using the term **area model** in place of *rug diagram*.)

You may need to help groups get started. One useful and familiar way to begin this analysis is to consider a large, convenient number of cases. The shooting probabilities are reported as percents, which can suggest imagining 100 cases. **Consider a large number of one-and-one situations, such as 100. In those 100, Terry would make 60 of her first shots. Can you show that in a rug diagram?** A diagram showing the outcomes after the first shot for 100 cases might look like this.

What does this rug diagram mean so far? What do you need to do next?

Some students may find it useful to work with graph paper, using a 10-by-10 section to represent 100 shots.

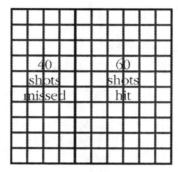

If they do use graph paper, be aware that later problems may not lend themselves so nicely to whole-number solutions, so students should also work with more schematic diagrams that encourage the computation of multistep probabilities by figuring area through multiplication and addition rather than by counting squares.

Some students may not know how to proceed, especially if they are thinking about each individual one-and-one situation. If so, you might ask what would happen in the cases in which Terry makes her first shot. **When Terry gets the opportunity to attempt a second shot, what portion of the time will she make this shot? How will you show this in your area model?** The result might look something like this.

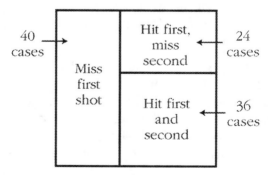

Students might develop and label this final area model in various ways. Here is one possibility.

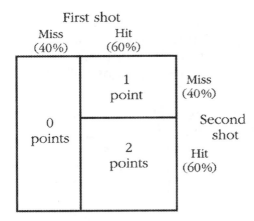

As you circulate, you may want to identify one or two groups whose members seem to have a clear understanding of the process and ask them to prepare presentations. In this case, presenters may be more able to share their thinking about how they designed their area models if they begin with unmarked rugs and talk through each decision in their analyses.

Discussing and Debriefing the Activity

Ask one or two groups to present their work by demonstrating each step of their analyses and how it gets incorporated into their area models. Emphasize the arithmetic of finding the portion of the total area for the various sections. For example, if students use a diagram like the one shown above, they will need to figure out that the "2 points" section represents 36% of the total area, because it is 60% of 60%.

As groups present their diagrams, they should note that each section contains both

- the number of cases (or portion of the area) that it represents
- the number of points scored for each case

If the diagram is drawn on a 10-by-10 grid, each box would represent a single one-and-one situation, and students would be able to find areas by counting boxes.

By considering both the number of cases and the point value for each section, students should come up with an analysis that is something like this, which is based on 100 cases:

40 cases worth 0 points each

24 cases worth 1 point each

36 cases worth 2 points each

Thus 100 cases give a total of

$(40 \cdot 0) + (24 \cdot 4) + (36 \cdot 2) = 96$ points

So, the average number of points per one-and-one situation is 0.96. This analysis also confirms that a score of 0 points is most likely (40% of the time), a score of 2 points is next most likely (36% of the time), and a score of 1 is least likely (24% of the time).

This analysis can be done without the use of an area model by simply analyzing what might happen in a large number of cases. But the combination of the geometric and arithmetic perspectives is generally helpful for students, and the technique of subdividing cases or area portions will be useful in analyzing the game of Pig.

This is a good time to reemphasize that the average result is the same no matter how many cases are considered. For instance, considering a total of 1000 one-and-one situations would give a total of 960 points, resulting in the same average of 0.96 point per one-and-one situation. You might mention that *expected average* might be a better term for this concept, but that *expected value* has become standard.

The rug diagram can help students confirm that the most likely score (which is 0) is not the expected value (which is very close to 1). The diagram might also provide insight into the misguided intuition that these values are the same, because some students can see in the diagram something "1-like" about the problem. Ask, **What's the difference between *expected value* and *most likely outcome*?** Stress that these two ideas are different, although often confused.

You might ask students to compare the one-and-one situation with other expected value problems they have looked at. One contrast to bring out is that in the one-and-one situation, the outcome did not happen all at once. In 60% of the cases, there was a second event to consider.

To keep students aware of the motivating unit problem, you might ask how this sort of analysis applies to the game of Pig. **What connections do you see in the theoretical analysis for one-and-one and our ideas about the game of Pig?** Because the number of rolls in Pig may consist of one roll or many, students can expect a "subdivided rug" to show up when they analyze that game.

Key Questions

Consider a large number of one-and-one situations, such as 100. In those 100, Terry would make 60 of her first shots. Can you show that in the area model?

When Terry gets the opportunity to attempt a second shot, what portion of the time will she make this shot? How will you show this in a rug diagram?

What's the difference between *expected value* and *most likely outcome*?

How does the one-and-one situation compare with other expected value problems you have looked at?

What connections do you see in the theoretical analysis for one-and-one and our ideas about the game of Pig?

Supplemental Activities

Free Throw Sammy (reinforcement) is a follow-up to *The Theory of One-and-One*. It could be used as a homework assignment instead of *Streak-Shooting Shelly* if you think that assignment is too difficult. This activity is simpler because Sammy's probability of making a shot does not change.

Which One When? (extension) brings out the idea that knowing the expected value for a situation does not always give all the information needed to make a decision.

Streak-Shooting Shelly

Intent
Students examine a variation on the one-and-one situation in which the probabilities change in the second stage. This continues their preparation for the analysis of the game of Little Pig.

Mathematics
This activity involves **conditional probability**, in which the probability of the outcome at one stage depends on the outcome of a previous stage. The mathematical goal is to deepen students' intuition about and analytic ability to work with probability rather than to derive formal algorithms or computation methods.

Progression
After individually exploring the multistage probabilistic situation posed, students will share approaches. The activity concludes with the naming of this type of problem as a question of conditional probability. Students will likely return to this topic in a few days when they consider *Martian Basketball*.

Approximate Time
20 minutes for activity (at home or in class)

15 minutes for discussion

Classroom Organization
Individuals, followed by whole-class discussion

Doing the Activity
Prior to assigning this activity, you might help students recall that their initial approach to analyzing the situation in *The Theory of One-and-One* was to consider a large number of trials and draw an area model.

Discussing and Debriefing the Activity
Ask for volunteers to present their results. Elicit at least one presentation involving area models. Area models provide a visual representation of what is happening and can help many students reason through multistage probability problems.

Students might use a sequence of diagrams to describe the stages of the situation. The following diagrams are based on a consideration of 100 one-and-one situations. This diagram represents the first shot.

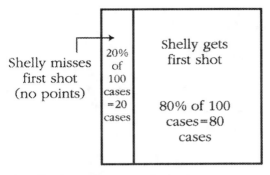

This diagram represents the first and second shot.

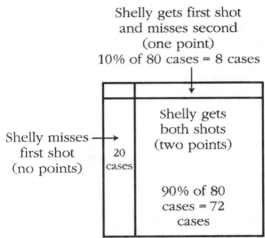

Thus, Shelly scores no points 20 times, one point 8 times, and two points 72 times. Confirm that 20 + 8 + 72 = 100.

In answer to Question 1, the diagram shows that she scores no points 20% of the time, one point 8% of the time, and 2 points 72% of the time.

For Question 2, students can use the diagram to figure total points. For 100 cases, Shelly scores a total of

$$20 \cdot 0 + 8 \cdot 1 + 7 \cdot 2 = 152 \text{ points}$$

yielding an expected value of 1.52 points per one-and-one situation.

Explain that this situation is an example of **conditional probability**, since the probability of Shelly making a free throw depends on when it comes in the shooting sequence.

Spins and Draws

Intent
This activity poses two more challenging expected value questions. Students will continue to develop methods to solve such problems and work through the meaning of expected value in various contexts.

Mathematics
Students find probabilities and expected values for a variety of situations. These problems become more challenging as the players in the "game" pay each other, and the context of a **fair game** is brought forth once again. Students continue to determine expected value through the "large number of trials" approach, and the large number has likely become a "convenient" number. Again, student reasoning and student-generated algorithms are the goal, rather than the derivation of a formula or method based on multiplying probabilities and points.

Progression
Students will work on this activity in small groups. Some students may draw area models to help solve the problems; others may be able to work through the tasks without using a visual representation by considering a large number of trials. Prior to moving onto Question 2, a more difficult situation, students should share ideas about how to make the game in Question 1 fair.

Approximate Time
35 minutes

Classroom Organization
Small groups, followed by whole-class discussion

Materials
The tasks in this activity can be challenging. Have paper clips (for making spinners) and decks of cards on hand in case students would like to use them to help them think about the problems.

Doing the Activity
You might engage students in the activity by introducing the first situation using an overhead spinner and asking students to guess who would come out ahead.

Tell students they are to work in their groups to solve the two problems in the activity, but that everyone must record and justify solutions. This may be a good opportunity to collect student work and assess understanding of the concept of expected value.

As students work, listen for their ability to explain their methods to one another. Prepare to ask questions during the class discussion based upon what you've heard.

Some students may neglect to consider the fact that Al and Betty pay each other. If a group fails to pick up on this, you might ask, **Where do the players get their money? What impact will that have on how much money they are left with after playing a large number of games?**

Groups who seem confident can be challenged to extend Question 1b with the question, **What other payment systems will make the game fair?** and even, **Can you describe all the payment systems that create a fair game?**

When groups have finished working on Question 1, bring everyone together for a class discussion.

Discussing and Debriefing the Activity

Have groups compare methods for tackling Question 1. Bring out both an area model approach and a "large number of trials" method. **If you used an area model, how did you divide the area? If you used a large number of trials, how many did you use?**

There are many ways to make the spinner game fair. One of the simplest is to change Al's payoff to $1.20, leaving Betty's payoff at 30¢.

Let groups return to Question 2. If time allows, discuss this question as well, which has some tricky aspects. For instance, since the charity receives a penny from each player if neither a jack nor a heart is drawn, the charity will get 2¢ each time it wins. Moreover, if the jack of hearts is drawn, both players have to pay and get paid, making a net gain of 12¢ for Ari. It might help to draw separate area models for Ari, Brenna, and the charity.

Key Questions

Where do the players get their money? What impact will that have on how much money they are left with after playing a large number of games?

If you used an area model, how did you divide the area? If you used a large number of trials, how many did you use?

What other payment systems will make the game fair?

Can you describe all payment systems that create a fair game?

Supplemental Activity

A Fair Dice Game? (reinforcement or extension) requires students to apply their knowledge of two-dice sums as well as the concept of a fair game, so it is a bit more difficult than either *A Fair Rug Game?* or the supplemental activity *Fair Spinners*.

Aunt Zena at the Fair

Intent
The activity involves another expected value exercise that, in the next activity, leads to simulating the problem on a graphing calculator.

Mathematics
In this expected value situation, there is a pay-to-play model involved, similar to many lotteries and insurance programs. The situation sets the context for students to think about a new technique for simulating a probabilistic situation: using a random number generator.

Progression
Students will work on the activity individually and then share ideas and solutions in a class discussion. Once students have clarified the key ideas in the questions and developed solutions, the teacher will introduce the idea of a random number generator in the following activity, *Simulating Zena*. Students then design and use their simulations to analyze the first situation posed in the activity experimentally.

Approximate Time
20 minutes for activity (at home or in class)

15 minutes for discussion

Classroom Organization
Individuals, then small groups, followed by whole-class discussion

Doing the Activity
Explain that this situation is a typical ring toss game at a fair that costs a set amount to play. Given the value of the prize and a certain player's chances of winning, students will analyze whether it is worth her while to play the game.

Discussing and Debriefing the Activity
Have students share their ideas in small groups. You might want to give two groups transparencies and pens to prepare a presentation for one of the two questions. While these groups prepare, you might challenge other groups to find Aunt Zena's expected value if she were to win once out of every five tosses (or some other variation on the game).

The "large number of trials" method is very helpful for this problem, but even with this method, students might have different approaches. Here are two possible approaches to Question 1.

If Aunt Zena plays 600 times, she would win about 30 times. One way to compute her net result is to multiply the $12 she wins on each successful toss by 30, for

$360, and balance this against the $600 she spent to play. This gives a net result of −$240.

Another perspective is to consider that each successful toss gives her a net gain of $11 and each unsuccessful toss represents a loss of $1. Her total is thus represented by the expression 30 • $11 + 570 • (-$1), or a net result of −$240.

In either case, Aunt Zena's expected value is −$240 ÷ 600 = −$0.40, or an average loss of 40¢ per toss.

Various approaches to Question 2 will show that Aunt Zena now has an expected gain of 20¢ per toss.

Simulating Zena

Intent
Students learn how to use a random number generator to conduct a simulation of a probability situation. They are being prepared for the end of *The Game of Pig* unit, where they will use a programmed simulation that relies on a random number generator.

Mathematics
The theoretical probability of an event can be estimated by running repeated simulations. Technology such as a graphing calculator can be very useful in running many simulations that depend on randomness. It might seem to be a contradiction to use a machine algorithm to produce numbers that are supposed to be unpredictable. So-called random number generators actually produce what are called *pseudorandom numbers*. They aren't random in the theoretical sense, but they are very close.

Progression
The teacher introduces the idea of using a random number generator to conduct a simulation of the situation presented in *Aunt Zena at the Fair*. Students then design and use simulations to analyze experimentally Question 1.

Approximate Time
15 minutes

Classroom Organization
Pairs, followed by whole-class discussion

Materials
Graphing calculators

Overhead graphing calculator

Doing the Activity
You might begin by asking the class, **How might you simulate the situation in Question 1 of *Aunt Zena at the Fair*?** Students might suggest something like putting 20 cubes in a bag, with one of them a different color from the others, and drawing out a cube to determine whether Zena wins or loses a given toss.

Explain that a graphing calculator has the capacity to do the same sort of thing. The calculator's random number generator will pick a random decimal between 0 and 1. If possible, demonstrate how this is done using an overhead calculator. Demonstrate many results of the random number generator so students can get a feel for what is occurring. You might ask, **What do you notice about what the calculator reports?**

Now turn back to the situation at hand. **How might you use a random number generator to simulate Aunt Zena's situation?** The most practical idea that students suggest will probably be something like, "If the number is .05 or less, she wins. If it's more than .05, she loses."

If students don't raise the question of what to do if the random number is *exactly* .05, you might bring it up yourself. Ask what they think makes sense. One approach is to decide arbitrarily whether .05 will count as a win or a loss. If students are concerned that either choice will be unfair, you can ask what the chances are of getting exactly .05, assuming that the random number generator is truly random and is picking 10-digit decimals. Help them to see that this is a 1-in-10-billion occurrence, so they shouldn't worry much about it.

Discussing and Debriefing the Activity

A class sample of about 600 trials (40 trials for each of 15 pairs of students) should give fairly reliable results. Combine the class results to see how close the fraction of simulated wins comes to the value of $\frac{1}{20}$ given in the problem.

Once these data have been collected, you might ask about the advantages of using a random number generator for conducting a simulation compared with using dice or colored cubes. **Do you prefer the concrete materials or the calculator?** Students should be able to see that this electronic technique, while not as tangible, is more easily adjusted to different situations than pulling cubes from a bag or rolling a die.

Key Questions

How might you simulate the situation in Question 1 of *Aunt Zena at the Fair*?

What do you notice about what the calculator reports?

How might you use a random number generator to simulate Aunt Zena's situation?

Do you prefer the concrete materials or the calculator?

The Lottery and Insurance—Why Play?

Intent
Students examine the role of expected value in two real-life situations.

Mathematics
Having a positive expected value is not always the only consideration in deciding whether to "play the game" in a real-life situation. In each of the situations in this activity, the expected return is less than the cost to "play." Lotteries and insurance are designed to make money for the companies administering them, so participants have to pay more to the companies than the average amount of money the company will pay out. So why do people play? In considering this question, students will incorporate cost-benefit analysis into their understanding of expected value.

Progression
Students will work on the activity individually and then share ideas and solutions in a class discussion.

Approximate Time
25 minutes for activity (at home or in class)

10 minutes for discussion

Classroom Organization
Individuals, followed by whole-class discussion

Doing the Activity
Students may need a brief introduction to the two questions. Encourage them to use mathematical vocabulary to explain their results.

Discussing and Debriefing the Activity
In the lottery example, spreading out the $6 million payoff over 14 million tickets gives an average payoff of 43¢ per ticket. A ticket costs $1, so the lottery player loses, on average, about 57¢ for each ticket bought. You might mention that most lotteries have lots of small payoffs as well as a big prize, but even so, the expected value must be less than the cost of a ticket for the lottery to stay in business. Similarly, for an insurance company to remain viable, its customers, collectively, must receive less from the company in insurance claims than they pay in premiums. In other words, on the average, people lose money when they buy insurance.

The question of "Why play?" can arise naturally from students or be asked by you. Students may have differences of opinion. Some may argue that for the lottery, the loss of $1 is of no consequence, while the gain of $6 million can radically change one's life, so people may think it is a worthwhile risk. Similarly, people might be able to afford insurance premiums but not the loss they might incur without insurance, so they "play the insurance game."

If your state or region has a lottery, students may want to learn more about how it works and where the proceeds go.

Ask students what the two situations in this activity have in common with *Aunt Zena at the Fair*. All three involve a "pay to play" circumstance, in which a player initially must spend money but may make back some of, all of, or more than that initial amount.

Key Questions

If the expected value in a lottery is negative, why would someone play?

Why would someone buy insurance?

Martian Basketball

Intent
This activity extends students' work with the two-stage events of the one-and-one problems to three-stage events.

Mathematics
Students find expected value in a three-stage event that includes the notion of **conditional probability.** In such a situation, creating rug diagrams is a challenge. Students will find that a tree diagram is another model for "seeing" how to determine expected value in multistage events.

Progression
At the conclusion of this activity, students will have three available approaches for determining the probabilities in a multistage event: area models, tree diagrams, and lists of combinations. While area models are the primary means for the analysis of Little Pig, tree diagrams can also be helpful.

Approximate Time
30 minutes for activity (at home or in class)

15 minutes for discussion

Classroom Organization
Individuals or groups, followed by whole-class discussion

Doing the Activity
Introduce the fictitious context of a one-and-one-and-one situation.

Discussing and Debriefing the Activity
Have students compare their answers. Then ask for presentations of students' solutions to the situations, including some that make use of area models and some that use tree diagrams. Some students may prefer to go straight to considering a convenient number of cases, but many will benefit from seeing area models, which will help them to visualize how each set of results is a fraction of a previous set of results.

Below are area models for Question 2. The diagrams are based on 1000 cases and show the situation after one and two shots, respectively. Students will benefit from attention to the steps by which the diagrams are derived.

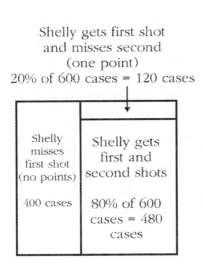

The final diagram might look like this.

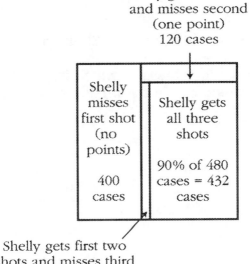

IMP Year 1, The Game of Pig Unit, Teacher Guide
© 2009 Interactive Mathematics Program

For 1000 trials, the results break down as follows:

3 points: 432 cases

2 points: 48 cases

1 point: 120 cases

0 points: 400 cases

This gives a total of 400 • 0 + 120 • 1 + 48 • 2 + 432 • 3 = 1512 points, for an expected value of 1.512 points per one-and-one-and-one situation. Students who base their analysis on a number of cases other than 1000 may well get nonwhole numbers. It's fine for them to round off to the nearest whole number, as long as they recognize that their answers are approximate.

This problem lends itself to introducing the idea of a tree diagram, which can be reinforced when discussing the next activity, *The Carrier's Payment-Plan Quandary*. The tree diagram can be developed in stages, one shot at a time. To introduce it, ask, **What are the possible outcomes for Shelly's first shot?**

The analysis can begin with a simple diagram showing what might happen with the first shot.

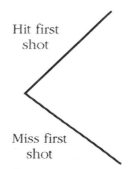

Students should be able to articulate that if Shelly misses the first shot, nothing happens, but if she makes the first shot, she takes another, which she either makes or misses. Gradually, the various ways in which the one-and-one-and-one situation can unfold will appear, leading to a diagram like this.

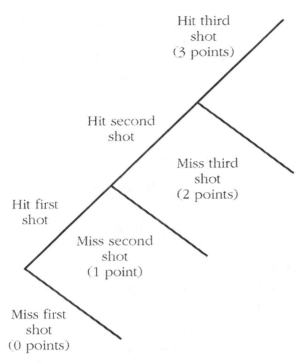

Introduce the term *tree diagram* for this figure and the term *branch* for each segment of the diagram. Branches that come off other branches are sometimes called *subbranches*.

The numerical information can be determined from a tree diagram, especially if it is used in coordination with the area models. In the next diagram, 1000 cases are portioned out at each stage to the proper branch.

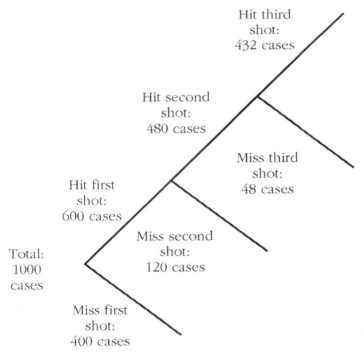

Note that in this unit, tree diagrams are used only to organize and picture outcomes. At this stage in the curriculum, finding probabilities for multistage outcomes by multiplying the probabilities along the branches is likely to lead to a mechanical application of the technique without real understanding.

Key Questions

Why might it be helpful to use 1,000 cases instead of 100 cases?

What different approaches did you use?

What are the possible outcomes for Shelly's first shot? For her second shot? For her third shot?

How can you represent each stage of Shelly's shot-taking experience?

Supplemental Activities

More Martian Basketball (extension) is a good follow-up to this assignment for students looking for a similar but more challenging problem.

Interruptions (extension) is a challenging problem that fits well here because students might use area models, tree diagrams, and simulations as tools for analyzing the situation.

The Carrier's Payment-Plan Quandary

Intent
Students will employ their available tools to compute an expected value and compare it to a simulated result.

Mathematics
In the typical payment plan, a customer pays the same amount each week or month for a given service. In this activity, a newspaper carrier is paid by a particular customer by choosing, each week, two bills from a bag containing five $1 bills and one $10 bill. The task is to determine the average weekly payment—that is, the expected value—and compare it to the results of a simulation.

Progression
The activities in *In the Long Run* bring together all of the ideas developed thus far and set students up for the activities in *Analyzing a Game of Chance*, in which they will return to the unit problem.

Approximate Time
5 minutes for introduction

25 minutes for activity (at home or in class)

30 minutes for discussion

Classroom Organization
Individuals, followed by whole-class discussion

Doing the Activity
You may want to brainstorm with students how they could carry out this simulation if they didn't have dollar bills—for example, by using five red cubes and one blue cube.

Discussing and Debriefing the Activity
Have students share their simulation methods and then compile the results. If the average of the data is close to the theoretical expected value of $5 per week, the class will probably not be able to draw any clear conclusions about which payment method produces more money for the newspaper carrier.

The class may have used a variety of approaches to find the theoretical probabilities of getting $2 and of getting $11. However students find the probabilities, the "large number of trials" method can then be used to determine the expected value. As students share how they found the probabilities, help them to focus on the distinction between *possible outcomes* and *equally likely outcomes*.

Labeling each individual bill can highlight the fact that the probability of drawing a $1 bill is not the same as the probability of drawing the $10 bill, but that the probability of drawing a particular $1 bill is the same as the probability of drawing the $10 bill. In other words, the equally likely outcomes for a given draw are the individual bills, not the amounts.

In the area models below, T represents the $10 bill, and O_1, O_2, O_3, O_4, and O_5, represent the five $1 bills. If you use subscript notation, you might mention how to read subscripted variables (for example, "oh sub one"). To create an area model, it may help to think of the bills as drawn one after the other, rather than both at once. After the first draw, the diagram could look like this.

First draw

| T | O_1 | O_2 | O_3 | O_4 | O_5 |

For each choice of the first bill, there are five possibilities for the second bill, so each column is divided into five sections, yielding 30 equal parts that represent the equally likely outcomes. The ten shaded boxes represent the cases in which a $10 bill and a $1 bill have been drawn, for a total of $11.

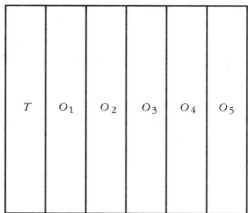

Second entry in each box gives second draw

IMP Year 1, The Game of Pig Unit, Teacher Guide
© 2009 Interactive Mathematics Program

Thus, $P(\$11) = \frac{10}{30} = \frac{1}{3}$ and $P(\$2) = \frac{20}{30} = \frac{2}{3}$.

The diagrams below avoid referring to individual bills.

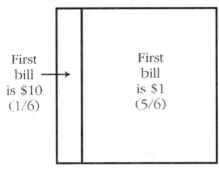

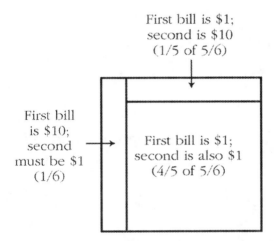

Some students may have constructed a tree diagram to represent the situation. If not, you might suggest this to reinforce the tool, which was introduced earlier. Such a tree diagram might look like this.

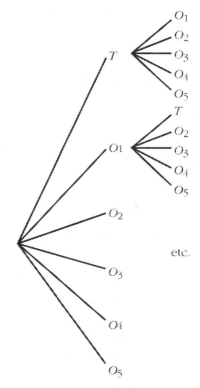

It may be less cumbersome to make a separate diagram for each set of subbranches.

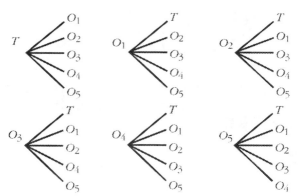

Altogether, there are 30 possible paths, which correspond to the 30 boxes in the area model. Because the paths are all equally likely, each path has a probability of $\frac{1}{30}$. Students can count paths to find the probability of each payment outcome.

Ask whether anyone used another approach. Some students may have listed the possible combinations, without regard to which bill was picked first. Such a list might look like this.

$TO_1, TO_2, TO_3, TO_4, TO_5$

$O_1, O_2, O_1, O_3, O_1, O_4, O_1, O_5$

$O_2, O_3, O_2, O_4, O_2, O_5$

O_3, O_4, O_3, O_5

O_4, O_5

As these 15 possibilities are equally likely, students can count to find the probabilities. You may want to ask why this method produces 15 possibilities, while the other methods show 30. The reason is that the list of combinations ignores different *orders*. For example, it shows TO_1 but does not also show O_1T.

However students determine the probabilities, they can use the "large number of trials" method to find the expected value. For example, for 300 trials, the carrier will get $11 about 100 times ($\frac{1}{3}$ of the cases) and $2 about 200 times ($\frac{2}{3}$ of the cases). Thus the carrier would expect to get $1,500 for 300 weeks, an expected value of exactly $5 a week. In other words, it turns out that the two plans have the same expected value.

Ask students, **Which payment plan would you choose, and why?** Although the expected value is the same for both plans, money may not be the deciding factor for some students as to which method is better. Some students may argue that the sense of adventure that accompanies the second plan makes it preferable even if it were to produce less money in the long run, while others may prefer the security of the first plan even if it were to earn less.

Key Questions

Can you make a tree diagram for this problem?

Did anyone use another approach?

Which payment plan would you choose, and why?

Supplemental Activity

Paying the Carrier (extension) asks students to devise a different payment system for the carrier.

A Fair Deal for the Carrier?

Intent
This simpler version of the carrier's payment-plan quandary is another real-life situation in which expected value may affect one's choice of options.

Mathematics
There are fewer outcomes in this activity, which involves a single-stage event, than in *The Carrier's Payment-Plan Quandary*. The simpler mathematics lends itself to a technology-based simulation.

Progression
Students have developed three methods for computing expected value: area models, tree diagrams, and listing all possible combinations. Now, whenever viable, they may also choose to use a random number generator to simulate a situation.

Approximate Time
5 minutes for introduction

20 minutes for activity (at home or in class)

25 minutes for discussion

Classroom Organization
Individuals, followed by whole-class discussion

Doing the Activity
Clarify the situation, and ask students to review the three methods they have been using to compute expected value (area models, tree diagrams, and listing all possible combinations). Remind students that whatever method they use, they will need to write a clear explanation of their solutions.

Discussing and Debriefing the Activity
Have students share their reasoning methods and answers. They will likely have computed the expected value using the "large number of trials" method. Save the result found for the expected value ($4.80) for later comparison with students' graphing calculator simulations in *Simulating the Carrier*.

You might ask for informal, rather than detailed, descriptions of students' simulation methods for Question 3.

Simulating the Carrier

Intent
Students create and use a technology-based simulation to gain insight into and confidence about their theoretical analyses.

Mathematics
This activity continues to reinforce students' understanding of "the average gain or loss in the long run" of a situation. Technology-based simulations give students the ability to play thousands of games in just minutes.

Progression
Students will use calculator or computer simulations to test and evaluate playing strategies for both Little Pig and Pig. The discussion here is based on the use of a graphing calculator for creating a simulation.

Approximate Time
25 minutes

Classroom Organization
Whole class, followed by groups

Materials
Graphing calculators

Doing the Activity
You may need to introduce the general idea of writing a program and using variables in programming. Here are three key points to bring out about what a calculator program is and how it works.

- It tells the calculator what to do and when to do it.
- It tells the calculator how to keep track of the numbers involved.
- It is written in a precise language, which varies from one machine to another.

You might introduce the metaphor of "labeled storage bins" for variables in a program. In order for the calculator to retrieve a number from a particular bin, the bin must have a label on it. Therefore, the program will have to tell the calculator how to label these bins as it uses them. Once a bin is labeled and has a number in it, students can tell the calculator to get the number from that bin or to replace that number with another.

The first stage in creating this simulation is determining how to use the random number generator to "draw a bill." For example, students might assign a number between 0 and .2 to represent drawing the $20 bill and a number between .2 and 1 to represent drawing a $1 bill. Whatever method they choose, ask them to explain how it corresponds to the probabilities of the problem. Before developing a simulation in which the calculator does all the record keeping, have students do a simulation in which the calculator simply chooses the bills and students record the results.

As an alternative to having students create their own programs, you might have them analyze an existing graphing calculator program or computer application file.

Discussing and Debriefing the Activity

With students working in groups, ask them to outline the entire simulation procedure repeated, say, 100 times. Here is one outline that might emerge.

Choose a random number.

If the random number is less than .2, add $20 to the total amount collected so far.

If the random number is more than .2, add $1 to the total amount collected so far.

When you have reached 100 trials, calculate the average of the results.

$$\text{average} = \frac{\text{total money received}}{100}$$

Ask groups to share their ideas about Questions 1 to 4.

Key Questions

What should the calculator do once it picks each random number?

Do the steps you set up have to be in a special order?

How can the program remember the random number after choosing it?

After 100 trials (or *n* trials), how will the calculator label what has happened?

How do the results of your simulation compare with the theoretical expected value?

Another Carrier Dilemma

Intent
Students use a variety of approaches to find the expected value in one more multi-step carrier situation.

Mathematics
Students continue to practice the three methods of calculating expected value (area model, tree diagram, and systematic list) in a multi-step situation.

Progression
This is an additional real-life situation for students to consider before returning to the unit problem.

Approximate Time
20 minutes for activity (at home or in class)

10 minutes for discussion

Classroom Organization
Individuals, followed by whole-class discussion

Doing the Activity
Clarify the point that students must use at least two methods to compute the expected value for the carrier's weekly earnings in this new version of the problem.

Discussing and Debriefing the Activity
Have at least one student present each method—an area model, a tree diagram, and a list of combinations—with a focus on finding the probability of each possible outcome. Finding the expected value once the probabilities are known should be fairly routine by now, and students should also understand that the expected value will be the same no matter which method they use.

You many want to share with students the area model below, in which all the bills are listed both across the top and down the side, and in which the impossible cases (representing the same bill being drawn twice) are shaded out. The number in each box shows the total amount the carrier earns.

	First draw				
Second draw	F_1	F_2	O_1	O_2	O_3
F_1		$10	$6	$6	$6
F_2	$10		$6	$6	$6
O_1	$6	$6		$2	$2
O_2	$6	$6	$2		$2
O_3	$6	$6	$2	$2	

Because the 20 possible cases are equally likely, one can determine by counting that the probability of getting $10 is $\frac{2}{20}$, the probability of getting $6 is $\frac{12}{20}$, and the probability of getting $2 is $\frac{6}{20}$. From these probabilities, the expected value over 100 weeks can be computed as $\frac{10(10) + 60(6) + 30(2)}{100}$, or $5.20 per week.

Key Question
Does one method for determining expected value seem more direct than another?

Analyzing a Game of Chance

Intent
In the activities in *Analyzing a Game of Chance,* students return to the unit problem to develop a strategy for playing the game of Pig.

Mathematics
Students begin *Analyzing a Game of Chance* with a valuable set of tools. They can find theoretical probabilities using area models, tree diagrams, and systematic lists; and they have ways of computing expected value. But virtually all of the situations they have encountered have consisted of a finite number of steps (usually one, two, or three flips, draws, or rolls) in each trial. In the game of Pig, a given trial could theoretically have many steps. To compute the expected value of a strategy that must take into account all of these possible outcomes, step-by-step methods will be ineffective and frustrating. To solve the unit problem, students must develop a new approach, based on assuming a particular state of the game and then calculating the expected value of risking one more step. They have the theoretical tools to do this but must learn to apply them in this new way.

To learn this new perspective, in *Analyzing a Game of Chance* students first consider two simplified versions of the game of Pig. Both variations include an infinite sample space, similar to Pig but with fewer possible outcomes. In these simplified contexts, students calculate complete expected values for finite strategies and develop the "one more step" analysis. Finally, they extend their insights to the more complex game of Pig.

Progression
The activities begin with an analysis of the game of Little Pig and proceed to applying the ideas learned to developing a complete solution—a complete strategy for maximizing the likelihood of winning—for "Big" Pig. The unit ends with assessments and the construction of the unit portfolio.

The Game of Little Pig

Pig Tails

Little Pig Strategies

Continued Little Pig Investigation

Should I Go On?

The Best Little Pig

Big Pig Meets Little Pig

The Pig and I

Beginning Portfolio Selection

The Game of Pig Portfolio

The Game of Little Pig

Intent
Examining this simplified game will support students in their analysis of the best strategies for playing "Big" Pig.

Mathematics
If students were to attempt to use area models and the concept of expected value to compare strategies for the game of Pig, they would find that the analysis gets complicated very quickly. Instead, they will now examine a game called Little Pig. This new game is simple enough that students can construct area models for many common strategies, but similar enough to Pig that it can suggest other ways to think about the best strategy for playing that more complicated game.

Progression
Work with Little Pig is exploratory. After analyzing many individual strategies for this game, students will consider how to find the best possible strategy. They will apply these insights later to find the best strategy for Pig itself without having to draw the complicated diagrams that quickly occur in an analysis of that game.

Approximate Time
35 minutes

Classroom Organization
Groups of 4, followed by whole-class discussion

Materials
Red, blue, and yellow cubes

Small paper bags

Sentence strips

Doing the Activity
As a class, read the rules of Little Pig. Give each group of students a bag containing one red, one yellow, and one blue cube. Just as students played Pig for the first time without much direction, have groups play this game, keeping track of their draws and scores for each turn. To help them analyze the game, ask them to keep track of both their own draws and scores and those of their partners.

Playing at least 10 rounds will be helpful. Students within each group should talk about the strategies they are using and be prepared to share one of those strategies with the class. Provide each group with a sentence strip for posting a strategy after playing the game.

Discussing and Debriefing the Activity

Be sure the strategies that students present are clear. If time allows, groups can play the game again to try out other groups' strategies.

Have students find the expected value for the strategy of stopping after the first cube is drawn no matter what the result. Working through this simple example may help with the next activity, *Pig Tails*.

Key Question

How is Little Pig like "Big" Pig? How is it different?

Pig Tails

Intent
This activity presents another simpler game to help prepare students to analyze Big Pig.

Mathematics
This variation of Pig uses coin flips. In *Pig Tails*, if you get tails, your turn is over. Students will evaluate strategies that include flipping only once each turn, flipping twice each turn, and so on, and then look for a pattern to determine the expected value per turn for flipping *n* times.

Progression
Students have another opportunity to analyze how the expected value increases or decreases with the number of flips per turn in a game similar to Pig.

Approximate Time
5 minutes for introduction

15 minutes for activity (at home or in class)

15 minutes for discussion

Classroom Organization
Individuals, followed by whole-class discussion

Doing the Activity
Clarify the rules for this game. Without simulating or even playing the game, students can move directly to considering a specific number of coin flips and calculating the expected value per turn. For each case, it will be helpful for them to draw a diagram or make a list. To develop a generalization, they will want to organize their expected value data and look for a pattern.

Discussing and Debriefing the Activity
Be sure everyone understands that for Question 1, following the one-flip strategy will score 1 point half the time and 0 points half the time, for an expected value per turn of $\frac{1}{2}$.

The key to Question 2 is recognizing that the only way to get a nonzero score is to get two heads in a row, and that this happens $\frac{1}{4}$ of the time, for a score of 2 points each time it happens. To help students see this, you might ask, **How many points do you get when you get a nonzero score?** Based on that probability, students should be able to see, perhaps using the "large number of games" approach, that the expected value per turn is again $\frac{1}{2}$.

For Question 3, you might need to focus the discussion on two ideas. **What is the probability of getting a nonzero score?** Students might have used an area model to see that the probability of getting three heads in a row is $\frac{1}{8}$. **How many points do you get when you get a nonzero score?** From the fact that this outcome scores 3 points, students should see that the average score per turn is $\frac{3}{8}$.

Discussion of the generalization (Question 4) is optional. For a strategy of flipping n times, the only way to get a nonzero score is to flip n heads. The probability of doing this is $\frac{1}{2^n}$, and the score for a sequence of n heads is n. Thus the expected value per turn for an n-flip strategy is $\frac{n}{2^n}$, which is $\frac{1}{2}$ if $n = 1$ or $n = 2$ and then decreases as n increases.

Key Questions

Is there a convenient way to keep track of the expected values for n flips? If you use a table, what labels might you use?

Can you see a pattern to help you generalize the expected value per turn for an n-flip strategy?

Supplemental Activities

Pig Tails Decision (extension) poses a strategy question for the game of Pig Tails in preparation for similar questions about Little Pig and Big Pig that students will soon encounter.

Get a Head! (extension) poses problems involving repeated flips of a coin and asks students to apply their developing understanding of probability and expected value. Question 2 asks them to analyze a game that could be arbitrarily long.

Little Pig Strategies

Intent
In this activity, students begin their analysis of the expected value per turn for the game of Little Pig.

Mathematics
In order to eventually analyze the best strategy for Big Pig, students continue to focus on the much less complicated game of Little Pig. Following the problem-solving strategy of "working with a simpler problem" will help students to make good choices when analyzing a more complicated situation.

Progression
Students continue to use an area model to find the expected value per turn of two Little Pig strategies: drawing until you get at least 2 points, or drawing two cubes.

Approximate Time
30 minutes

Classroom Organization
Groups or pairs, followed by whole-class discussion

Doing the Activity
If time is a concern, you may want to have some groups start on the 2-draw strategy and others on the 2-point strategy and then have them share the area models they create.

All groups should find the expected value per turn for both strategies so they can then compare the strategies.

Discussing and Debriefing the Activity
As strategies are evaluated, have groups post their area models and expected values on a chart for comparison purposes later.

Ask, **What steps did you use in developing your area models?**

You might start presentations with the 2-point strategy, as the associated area model is slightly simpler. Ask presenters how their groups developed the diagrams. They might begin with the first diagram below. Because scoring 1 point is the only outcome that requires a second draw, the middle column can be subdivided to produce the second diagram.

| yellow: 0 points | red: 1 point | blue: 4 points |

yellow: 0 points	red, then yellow: 0 points	blue: 4 points
	red, then red: 2 points	
	red, then blue: 5 points	

A key aspect of the problem is realizing that the four possible scores—0 points, 2 points, 4 points, and 5 points—are not equally likely, nor do the five boxes represent equally likely outcomes. To get at this realization, you might ask, **What is the probability of each outcome?** Subdividing each of the outer columns may help students recognize each of the smaller boxes in the middle column is $\frac{1}{9}$ of the total area, giving these probabilities.

$$P(0 \text{ points}) = \frac{1}{3} + \frac{1}{9} = \frac{4}{9}$$

$$P(2 \text{ points}) = \frac{1}{9}$$

$$P(4 \text{ points}) = \frac{1}{3}$$

$$P(5 \text{ points}) = \frac{1}{9}$$

Ask, **Have you solved the problem?**

Students may be so engaged in finding probabilities that they forget to compute the expected value. Using the "large number of trials" approach with 900 trials, the cases break down like this.

Yellow on first draw: 300 cases

Red then yellow: 100 cases

Red then red: 100 cases

Red then blue: 100 cases

Blue on first draw: 300 cases

For ease in finding the expected value of $2\frac{1}{9}$, which is about 2.11, students may prefer to list the cases by score rather than color sequence.

0 points: 400 cases

2 points: 100 cases

4 points: 300 cases

5 points: 100 cases

Posting this expected value and an area model to explain the result will allow comparisons with strategies and the simulation to be discussed later. The area models will be useful because gaining insight into strategies for Pig will require asking why one strategy results in an expected value that is larger than another.

The analysis for the 2-draw strategy is similar. For this area model, the section representing a blue cube on the first draw must also be subdivided.

	red, then yellow: 0 points	blue, then yellow: 0 points
yellow: 0 points	red, then red: 2 points	blue, then red: 5 points
	red, then blue: 5 points	blue, then blue: 8 points

Using 900 cases gives this breakdown.

0 points: 500 cases

2 points: 100 cases

5 points: 200 cases

8 points: 100 cases

Post the diagram and the expected value of $2\frac{2}{9}$, which is approximately 2.22.

Ask, **Why is the 2-draw strategy better?** Students can see that the expected value of the 2-draw strategy is higher. It's good to pursue the question of why and to ask where the difference in the strategies shows up in the analysis.

Bring out the idea that the only difference in play occurs if the first cube is blue. With the 2-point strategy, the player stops; with the 2-draw strategy, the player draws again. In terms of the 900-case analysis, drawing again replaces the 300 4-point games (1200 points) with 100 0-point games, 100 5-point games, and 100 8-point games (1300 points). Thus if you draw blue on the first cube, you are better off drawing again than stopping.

Key Questions

What steps did you use in developing your area model?

What is the probability of each outcome?

Have you solved the problem?

Why is the 2-draw strategy better?

Continued Little Pig Investigation

Intent
This activity gives students additional opportunities to evaluate strategies for Little Pig.

Mathematics
Students make an organized list of "point" strategies and "draw" strategies for Little Pig.

Progression
This activity is written very generally to allow students to progress at their own pace. As more information is added to the chart of expected values for Little Pig strategies, students will find it easier to make decisions about how to determine the best strategy for playing the game.

Approximate Time
100 minutes over 2 or 3 days

Classroom Organization
Individuals, pairs, or groups, followed by whole-class discussion

Doing the Activity
As students work, continue to post expected values and area models.

Some students will begin with 3-point and 3-draw strategies; others will move on to 4-point and 4-draw strategies. Some may want to analyze a strategy other than a "point" or "draw" strategy.

Discussing and Debriefing the Activity
As a class, compare the 3-point strategy to the 2-point strategy, and compare the 3-draw strategy to the 2-draw strategy. Analyze the 4-point strategy and other strategies as time allows. Post area models and expected values for all new strategies, organizing the point and draw strategies on the chart.

The 3-point strategy, with an expected value of $2\frac{2}{9}$, can be represented by this area model, which differs from the model for the 2-point strategy only in that the "red then red" section is further subdivided.

yellow: 0 points	red, then yellow: 0 points	blue: 4 points
	r, r, y: 0	
	r, r, r: 3	
	r, r, b: 6	
	red, then blue: 5 points	

The 3-draw strategy, also with an expected value of $2\frac{2}{9}$, can be shown with a diagram like that below.

yellow: 0 points	red, then yellow: 0 points	blue, then yellow: 0 points
	r, r, y: 0	b, r, y: 0
	r, r, r: 3	b, r, r: 6
	r, r, b: 6	b, r, b: 9
	r, b, y: 0	b, b, y: 0
	r, b, r: 6	b, b, r: 9
	r, b, b: 9	b, b, b: 12

These two diagrams can help students compare the two strategies and see why they yield the same expected value. Although subdividing the 2-draw strategy's "red then red" box increases the value from 2 points to an average of 3 points, subdividing its "blue then blue" box decreases the value from 8 points to an average of 7 points.

It may be of help to have the class describe the two types of strategies discussed so far.

- Point strategies are those in which you stop when you reach a preset number of points.
- Draw strategies are those in which you stop after a preset number of draws.

If students analyzed strategies in addition to these two types, have them present those results and compare them with the other strategies considered so far.

Use your judgment about how long to have students continue analyzing strategies. At some point, you may want to interrupt and simply give them the expected values for additional point and draw strategies (listed below). Your goal is to have them realize, for use in analyzing Pig, that except for the "small" cases, point strategies are superior to draw strategies, as illustrated by the expected values.

Strategy	Expected value
2-point strategy	$2\frac{710}{2187} \approx 2.325$
2-draw strategy	$2\frac{2}{9} \approx 2.22$
3-point strategy	$2\frac{2}{9} \approx 2.22$
3-draw strategy	$2\frac{2}{9} \approx 2.22$
4-point strategy	$2\frac{20}{81} \approx 2.247$
4-draw strategy	$1\frac{79}{81} \approx 1.975$
5-point strategy	$2\frac{88}{243} \approx 2.362$
5-draw strategy	$1\frac{157}{243} \approx 1.646$
6-point strategy	$2\frac{88}{243} \approx 2.362$
6-draw strategy	$1\frac{231}{729} \approx 1.317$
7-point strategy	$2\frac{710}{2187} \approx 2.325$

Given these results, students will likely notice that the draw strategies are getting worse as the number of draws increases, which may suggest to them that one of the point strategies is best. They may also see that the 7-point strategy is not as good as the 6-point strategy, which suggests that continuing to draw beyond 7 points is not worthwhile, and that the 5-point and 6-point strategies both seem to yield the maximum possible expected value.

Key Questions

What types of strategies have we seen?

Why do the 3-point and 3-draw strategies have the same expected value?

What appears to be the best strategy?

Should I Go On?

Intent
Students connect their investigation of strategies for playing Little Pig to the game of Big Pig.

Mathematics
This activity shifts the focus from calculating expected value of a point or a draw strategy to deciding whether it is beneficial to draw again based on the points a player has already accumulated. Calculating a new expected value will answer the question, "Given what I have now, would I benefit in the long run from drawing again?"

Progression
Following this activity, students will be ready to revisit Big Pig and determine a best strategy for playing it.

Approximate Time
5 minutes for introduction

20 minutes for activity (at home or in class)

15 minutes for discussion

Classroom Organization
Individual or pairs, followed by whole-class discussion

Doing the Activity
Clarify for students that in this activity the focus is on deciding, in each of two situations, whether a player should draw again (to increase the expected value) or stop with the current points (because an additional draw will not increase the expected value). In other words, the focus is not on how students arrive at a particular point score, but rather on whether they should draw again. This idea connects to the gambler's fallacy and the belief of some players that yellow is "due" to be drawn. It is important for students to recognize that regardless of a player's current score, the next draw still has an equal chance of being yellow, red, or blue.

Discussing and Debriefing the Activity
In Question 1, for the class in which students draw one more time, one would expect about ten students (out of 30) to draw yellow and end the turn with 0 points, ten to draw red and end with 11 points, and ten to draw blue and end with 14 points. This gives a total of 250 points, for an average of about 8.33 points per student. Thus this class ends up worse off, on average, than the class in which students stop at 10 points.

Take note that both classes begin with averages much higher than the expected value for any of the Little Pig strategies that students have examined. That is because all the students in this problem start with 10 points. Therefore, if you get to 10 points, you will be better off *in the long run* if you stop rather than draw again.

In contrast, in Question 2 the class that draws again raises the average from 2 to 3 points per student. So, if you have only 2 points, you will be better off *in the long run* if you draw again rather than stop.

Key Question
What conclusions can you draw about Little Pig from this activity?

The Best Little Pig

Intent
In this activity, students will discover a clear pattern that they can use to determine how to play Little Pig confidently.

Mathematics
Using the results from *Should I Go On?* students identify the best strategy for Little Pig. There are points in the game at which drawing again will increase the expected value, and there are points at which drawing again will decrease the expected value. Finding the *break-even point* will lead to the optimal strategy for playing the game.

Progression
This final Little Pig investigation will prepare students to return to the unit problem: finding the best strategy for playing Big Pig.

Approximate Time
25 minutes

Classroom Organization
Groups, followed by whole-class discussion

Doing the Activity
This activity asks students to identify the best strategy for Little Pig. If students are not sure how to proceed, you might suggest that they apply their reasoning from *Should I Go On?* to some cases between 0 and 10 points.

Discussing and Debriefing the Activity
For the discussion, develop a chart of values in which the *In* column represents current points and the *Out* column represents potential points *in the long run* if one draws again.

Have students share their discoveries. They should see that for initial scores of less than 5 points, a player is better off drawing again, but with scores of more than 5 points, a player is better off stopping. With exactly 5 points, the expected value of the next draw is 0, so mathematically it doesn't matter whether the player draws once more or not—the choice at this point is psychological, not mathematical.

Students may wonder about the 6-point strategy, which has the same expected value as the 5-point strategy. If so, help them see that waiting for 6 or more points before stopping should have the same effect as waiting for 5 or more, but in the 6-point case the player *must* stop to maximize the score in the long run, whereas in the 5-point case the player has the option of continuing for one more draw. Thus either the 5-point or the 6-point strategy is optimal for Little Pig.

At some point, you might ask, **Why do you think there is a point at which an additional draw gives no gain?**

To bring out the idea of a transition value, you might encourage use of the problem-solving technique of considering extreme cases. Students will probably recognize, for instance, that if you have 0 points, it's better to draw than to stop. (You have nothing to lose, and you might gain.) On the other hand, if you have a very high score (such as 100 points), you have a lot to lose, compared to how much you might gain, so you should stop. Somewhere in between—in this game, at 5 points—the situation switches from "better to go on" to "better to stop."

Identify this transition value as the *break-even point*.

Using technology to simulate the results of strategies can add to student understanding. You might use some of these programming ideas.

Key Question
Why do you think there is a point at which an additional draw gives no gain?

Big Pig Meets Little Pig

Intent
Having developed an approach to analyze Little Pig, students now return to the unit problem.

Mathematics
Many people initially think that the best strategy for Pig is to stop after a particular number of rolls. The analysis in the unit shows that, in fact, the best strategy is to stop after reaching a particular number of points.

Progression
Students apply their insights from finding the best strategy for Little Pig to the unit problem for Big Pig.

Approximate Time
10 minutes for introduction

25 minutes for activity (at home or in class)

40 minutes for discussion

Classroom Organization
Individuals, followed by whole-class discussion

Doing the Activity
Review the rules for the original game of Pig.

Discussing and Debriefing the Activity
Although the computations are more cumbersome, the basic idea here is identical to that in *Should I Go On?*

For example, if there are 30 students with 10 points each, and they each roll once more, one would expect five students to end the turn with 12 points, five to end with 13 points, five to end with 14 points, five to end with 15 points, five to end with 16 points, and five to end with 0 points. This is a total of 350 points, for an average of 11.67 points per player.

By this analysis, a player with 10 points is better off in the long run rolling again.

It is likely that some students will have found that the break-even point for Big Pig is 20 points. That is, a player with fewer than 20 points is better off rolling again, a player with more than 20 points is better off stopping, and for a player with exactly 20 points, the two options are equally good. If no one has reached this conclusion, the class might continue to work until they do so, perhaps with different groups testing different numbers.

This analysis is the culmination of the unit's work. It answers the question posed at the beginning of the unit, so it should receive appropriate fanfare.

Students can confirm the results so far by using a calculator or computer simulation. A simulation of 200,000 or more games should be enough to indicate that the best strategy (of those that can be tested) is one in which a player stops after *about* 20 points. But there's an amazing amount of variation from one run to the next. So it takes more games than you want to do to make a clear distinction as to *exactly* what the very best stopping point is.

You might want to try a large number of turns with both the 20-point and 21-point strategies. If time allows, try 19 or 22 points to see what happens. (Unfortunately, students will not be able to test mixtures of "roll" and "point" strategies with these simulations.)

Supplemental Activities

Pig Strategies by Algebra (extension) asks students to state strategies for Little Pig and Big Pig algebraically, with point values expressed as variables.

Fast Pig (extension) uses a variant of the game of Pig to lead students toward a general analysis of the expected value for an *n*-roll strategy for Pig.

The Pig and I

Intent
Students synthesize their work on the best possible strategy for the game of Pig.

Mathematics
Students justify their decisions, based on the three models for expected value, and describe "roll" versus "point" strategies for the game of Pig.

Progression
Students formalize their own conclusions about the best way to play the game of Pig.

Approximate Time
10 minutes for introduction

20 minutes for activity (at home or in class)

Classroom Organization
Individuals, followed by whole-class discussion

Doing the Activity
Explain that students will now write a persuasive essay, supported by the mathematics of probability. Ask students to share some ideas about what they might write about, such as what they learned about various strategies and how the strategies differ.

Discussing and Debriefing the Activity
You might ask a few students to read their summary comments to the class.

Beginning Portfolio Selection

Intent
Students review their work and select activities that represent simulations or experiments and theoretical analyses.

Mathematics
This unit develops two main approaches for finding probabilities. One is the use of simulations and experiments to generate data and construct observed probabilities. The other is the use of area models, tree diagrams, organized lists, and other methods to calculate theoretical probabilities. These ideas are the focus of this initial portfolio selection.

Progression
Students explain the activities they have selected and describe what they have learned about probability from them.

Approximate Time
20 minutes

Classroom Organization
Individuals, followed by whole-class discussion

Doing the Activity
Before students begin their writing, it may help to brainstorm a list of activities and mathematical topics studied through the unit.

Discussing and Debriefing the Activity
You may want to have a few students read their descriptions to the class.

The Game of Pig Portfolio

Intent
Students synthesize their work on the game of Pig by assembling portfolios and writing cover letters.

Mathematics
Students describe the central unit problem and the main mathematical ideas explored in the unit. Their descriptions should give an overview of how the key ideas were developed and how they were used to solve the central problem.

Progression
This activity concludes the unit as students reflect on and summarize the mathematical ideas they have encountered.

Estimated Time
30 minutes

Classroom Organization
Individuals

Doing the Activity
Remind students that their portfolios have three parts: a cover letter that summarizes the unit, papers they have selected to include from their work in the unit, and a discussion of their personal growth during the unit.

Rug Games

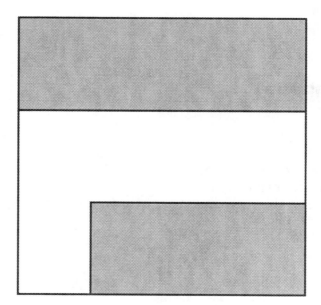

Rug Games (continued)

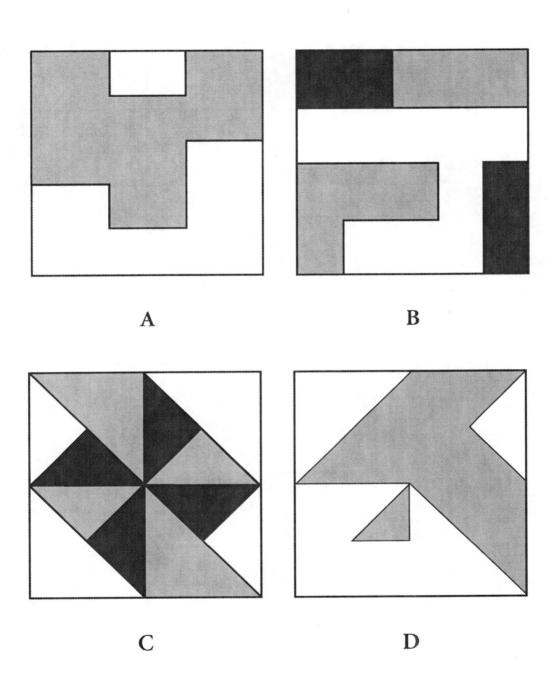

A

B

C

D

1-Centimeter Graph Paper

1-Inch Graph Paper

$\frac{1}{4}$-Inch Graph Paper

In-Class Assessment

Consider the following game.

In each turn of the game, you flip a coin three times.

If you get three tails, you win 9 points.

If you get the sequence "tail, tail, head," you get 3 points.

If you get any other sequence, you get no points for that turn.

What is your expected value per turn for this game? Explain your reasoning.

Take-Home Assessment

1. Tetrahedron Dice

Suppose you have two dice shaped like tetrahedrons. (A tetrahedron is a solid figure with four faces.)

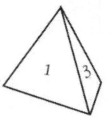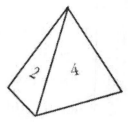

These dice have the numbers 1, 2, 3, and 4 on them, one number on each face. When one of these dice is rolled, each face is equally likely to end up on the bottom. The number on the bottom is considered the result when you roll one of these dice.

Make up a game for two players in which the winner is determined by the sum of the results from rolling this pair of dice. Create your game so that one player has a probability of 7/16 of winning and the other player has a probability of 9/16 of winning.

Explain the rules clearly, and explain how you know what each probability is.

2. County Fair

You are setting up a booth at the county fair. You will have a paper bag containing one $10 bill, two $5 bills, and two $1 bills. Each contestant will draw two bills out of the bag. If the contestant draws exactly $15 or exactly $10, he or she gets to keep the bills. Otherwise, the contestant gets nothing.

If you charge $3 for each contestant, will you make money or lose money in the long run? How much will you gain or lose, on average, for each contestant? Explain your answer.

The Game of Pig Guide for the TI-83/84 Family of Calculators

This guide gives suggestions for selected activities of the Year 1 unit *The Game of Pig*. The notes that you download contain specific calculator instructions that you might copy for your students. The download also includes programs that you can download from your computer to calculators. NOTE: If your students have the TI-Nspire handheld, they can attach the TI-84 Plus Keypad (from Texas Instruments) and use the calculator notes for the TI-83/84.

The Game of Pig unit explores basic principles of probability, including the concept of expected value. You will find the graphing calculator's random number generator useful for simulating random events. This feature, combined with the calculator's programming capability, allows students to create and run simulations of games. Because technology provides a faster method for repeating a procedure, students can use technology to run large numbers of trials quickly.

You can use the calculator to simulate rolling a die or picking randomly from a deck of cards. If you want to avoid using dice or cards with students, you might prefer these calculator simulations.

You'll notice that simply entering the steps of a premade program into a calculator does not give you much understanding of the program. If you would like students to develop more understanding of programming, have them write their own programs from scratch or modify the premade ones included here.

The Game of Pig: This is the first day students play the game of Pig. You can replace the die rolling at any point in the unit with a simple calculator simulation that uses random integers. The simulation generates results more quickly than rolling a die and has the added advantage that you can see your previous results as you generate new ones. *The Game of Pig Notes for the TI-83/84 Family of Calculators* has a detailed discussion in "Random Numbers on the Calculator."

Waiting for a Double: In this activity, students roll two dice at once. "Random Numbers on the Calculator" discusses how to use random integers for a simple simulation. The note "Simulating Waiting for a Double" describes an optional calculator program that can be used to simulate an entire class's data and calculate the average number of rolls before getting doubles.

The Gambler's Fallacy: One way to get more data is to pool the data from several classes. Another way is to use the optional program in "Simulating the Gambler's Fallacy," which can quickly repeat the students' experiment.

Expecting the Unexpected: This activity introduces frequency bar graphs. You will probably want students to have experience making these graphs by hand. However, after ample exposure to hand-drawn graphs, calculator-generated frequency bar graphs can be used to represent individual or class results. See the information in "Making a Frequency Bar Graph."

Rollin' Rollin' Rollin': Again, calculator-generated frequency bar graphs can be used to represent individual or class results. See the notes in "Making a Frequency Bar Graph." The lists can also be defined with random integer commands, to create a simulated bar graph.

Spinner Give and Take: In this activity, students can simulate the spinner on their calculators. See "Random Numbers on the Calculator." Once again, you'll find that the calculator is a more efficient tool for creating many random events quickly; unlike a handmade spinner, it is not influenced by variables such as the tilt of the desk or the quality of the cardboard. However, the spinner has the advantage of being something students can build and use themselves.

Mia's Cards: In this activity, students work with the probability of pulling random cards from a deck. If you have students interested in programming simulations on the calculator or you want to avoid using cards with students, have them write a program to draw a random card from a standard deck. "Pick a Card" gives instructions for such a program.

One-and-One: Students can program their calculators as a simulation of this activity. You can use "Simulating One-and-One" either to replace or to supplement the simulation in which students pick colored cubes from a bag. Students can also use this simulation when they complete *A Sixty-Percent Solution.*

Simulating Zena: If you have not used simulations on the previous activities, you can introduce random number generators now. "Random Numbers on the Calculator" describes the random number generator on a calculator in more depth. It is appropriate for students who have never used this feature before, as well as those who are already somewhat familiar with it.

Martian Basketball: Students who would like a programming challenge might enjoy programming a simulation for this activity. They can use "Simulating One-and-One" as a model to help them get started.

Simulating the Carrier: Students who have outlined a program in *A Fair Deal for the Carrier?* might use those ideas as they plan a program for *Simulating the Carrier.* You will want students to share ideas about the program outline and develop a general plan. But at some point you and your students will need to turn this plan into an actual program. Depending on how much programming your students have already done, you may need to introduce new programming ideas. "Programming the Carrier Simulation" gives a sample program for simulating the carrier situation from *The Carrier's Payment Plan Quandary* and can be used at this point. If you are pressed for time, you may want to present students with a finished program rather than have them create the program on their own. You can use the calculator's linking capability to pass the program from calculator to calculator.

The Game of Little Pig: In this game, students pick one of three random cubes from a bag. Students can simulate 1-in-3 probability using random integers on their calculators. They can also use random decimals and determine each result by deciding whether the number is less than 0.33333..., more than 0.6666..., or between these two numbers.

Big Pig Meets Little Pig: "A Simulation for the Game of Pig" contains the program for simulating both versions of the game of Pig. Once the program is entered in one calculator, students can share the program by linking their calculators. If you have students interested or skilled in programming, ask them to write a simulation of Pig themselves.

The Probability Simulation Application

In addition to the calculator commands and programs described in *The Game of Pig Notes for the TI-83/84 Family of Calculators*, you may also find the calculator application *Probability Simulation* useful. This free application is distributed by Texas Instruments (visit education.ti.com) and comes preloaded on most TI-83 Plus and TI-84 Plus calculators. (Applications cannot be loaded onto a standard TI-83.) To see if a calculator has *Probability Simulation* installed, press APPS and look for **Prob Sim**.

Probability Simulation can be used to simulate dice, coins, spinners, marbles, and cards. Some of the simulations allow you to collect the results of several trials and make frequency bar graphs of the results. The application uses simple animation, so it is a nice transition from physical, hands-on simulation to abstract, random-number-generator simulation. However, the animation makes some simulations work slower than a comparable text-based program.

 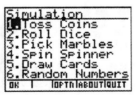

If you choose to use this application, visit education.ti.com and download the *Probability Simulation User's Manual*, which documents all of its features and options.

Random Numbers on the Calculator

Many probability problems involve the idea of picking something at random. In the simplest cases this means that every possible "pick" is equally likely.

Your calculator has features called random number generators. You can use these to simulate probability experiments.

The **rand** command generates a random number between 0 and 1. To access this command, press [MATH], highlight **PRB**, choose **rand**, and press [ENTER].

The instruction **rand** should appear on your home screen. Press [ENTER] again, and the calculator will display a decimal between 0 and 1. Press [ENTER] more times to generate more random numbers.

Simulating Flipping a Coin

A common use of random numbers is to simulate probability experiments. For example, you can simulate flipping a coin by letting a random number from 0 to 0.5 represent heads and random numbers from 0.5 to 1 represent tails.

Simulating Rolling a Die

You can use a similar method to simulate rolling a six-sided die. Let random numbers between 0 and $\frac{1}{6}$ (from 0 to 0.1666666666) represent rolling a 1, random numbers between $\frac{1}{6}$ and $\frac{2}{6}$ (from 0.1666666667 to 0.3333333333) represent rolling a 2, and so on.

Continued on next page

The Game of Pig

IMP Year I: Calculator Notes for the TI-83/84 Family of Calculators

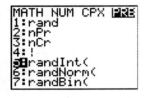

Generating Random Integers

The **randInt(** command generates a random integer within a specified range. You'll find **randInt(** further down in the [MATH] **PRB** menu.

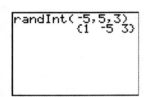

After entering this command, enter lower and upper bounds for your range of integers (separated by a comma) and a closed parenthesis. Before you close the parentheses, you can also enter a third number that represents the number of trials.

Simulating Rolling a Die with Integers

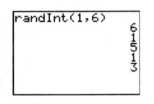

To simulate rolling a single six-sided die, enter **randInt(1,6)**. Keep pressing [ENTER] to simulate multiple rolls. Here are the results of five rolls.

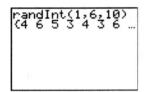

You can also simulate multiple rolls by entering a number of trials. This screen shows the result of ten rolls of a single die. The ellipsis (...) means that you can use the left and right arrow keys to see the rest of the data.

Simulating Rolling Two Dice with Integers

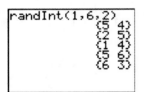

To simulate rolling two dice at once, enter **randInt(1,6,2)**. Keep pressing [ENTER] to simulate multiple rolls of the pair.

Simulating Flipping a Coin with Integers

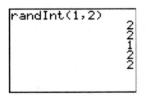

To simulate flipping a coin, enter **randInt(1,2)** and let 1 represent heads and 2 represent tails. The results here represent tails, tails, heads, tails, tails.

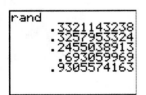

An alternative is to enter **rand** and let numbers less than 0.5 represent heads and numbers greater than 0.5 represent tails. This screen shows heads, heads, heads, tails, tails.

Continued on next page

Simulating Other Probabilities

You can use **rand** to model situations in which the outcomes are not equally likely.

For example, suppose you want to simulate spinning a spinner that has one quarter-section shaded and the rest white. The probability of the spinner stopping in the shaded region is $\frac{1}{4}$. So, you can let random numbers between 0 and 0.25 represent the shaded region; numbers between 0.25 and 1 will represent the white region.

Recognizing That the Random Number Generator Isn't Truly Random

Unfortunately, there's no such thing as a perfect random number generator. Your TI calculator uses a complex mathematical process to produce predetermined sequences of numbers as output for the **rand** and **randInt** commands. For most simulations the calculator's random number generators work quite adequately.

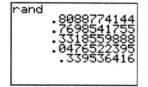

You can observe this lack of randomness when your calculator is brand-new, because new calculators start at the same place in the random number sequence. This means that if two people with new calculators both use the **rand** command, they should get exactly the same numbers.

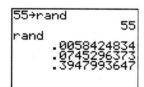

You can also observe how this sequence is not random using calculators that are not new. To do this, work with a partner who also has a calculator. Agree on a whole number, say, 55. You will access the fifty-fifth number in the random number sequence by first entering **55→rand** on each calculator. (The [STO▶] key gives you the → symbol.) Press [ENTER].

Now enter the **rand** command again. Compare your results with those of your partner.

Keep pressing [ENTER] to generate more "random" numbers. Compare your results with your partner. What do you notice?

IMP Year 1: Calculator Notes for the TI-83/84 Family of Calculators   The Game of Pig

Simulating Waiting for a Double

You have learned how to use the command **randInt(1,6)** to simulate rolling a single die. The program below uses this command twice to count how many times it takes to roll a double with two dice. It can repeat this process for as many trials as you want, and then it will calculate the average number of rolls needed for a double.

The column on the left shows you each line of the program, and the column on the right gives some explanation of the command.

Instruction	Explanation
PROGRAM:DOUBLES	Press [PRGM], select the **NEW** menu, and press [ENTER]. Name the program **DOUBLES** and press [ENTER].
:ClrHome	Clears the home screen. Press [PRGM], select the **I/O** menu, and scroll down to **8:ClrHome**.
:ClrList L1	Clears any data in list 1. Press [STAT] and select the **EDIT** menu for **ClrList**. Then press [2nd] [L1].
:Input "TRIALS?",T	Asks how many trials of the experiment to perform; saves that value as T. Press [PRGM] and select the **I/O** menu for **Input**.
:For(N,1,T)	Repeats the experiment N times, where N goes from 1 to T. Press [PRGM] and select the **CTL** menu for **For(**.
:0→I	Variable I counts the number of rolls needed for a double on each trial of the experiment. Use the [STO▶] key to store 0 as the starting value. Each time the experiment is repeated, I is set back to zero.
:Repeat F=S :randInt(1,6)→F :randInt(1,6)→S :I+1→I :End	These five lines repeatedly roll two dice (F for first and S for second) until the outcomes on both are equal. Each time they are rolled, variable I is increased by 1. Press [PRGM] and select the **CTL** menu for **Repeat** and **End**.

Continued on next page

: 1→L1(N)   The number of rolls needed for a double is stored as the Nth piece of data in list L1.

: End   After N reaches the value of T, the For command ends.

: Disp "AVG ROLLS:", mean(L1)   The program finds the mean (average) number of rolls needed for a double. Press [PRGM] and select the I/O menu for Disp. Press [2nd] [LIST] and select the MATH menu for mean(.

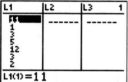

By using the program to do a large number of trials, you can get good results much faster than rolling dice by hand.

If necessary, after you run the program you can see the number of rolls needed on each individual trial by going into the list editor. Press [STAT], select the EDIT menu, and choose Edit.

IMP Year 1: Calculator Notes for the TI-83/84 Family of Calculators   **The Game of Pig**

Simulating the Gambler's Fallacy

This program allows you to quickly repeat the experiment in *The Gambler's Fallacy*. You can do a large number of trials and the program will count the number of triplets followed by the same or a different outcome.

The column on the left shows you each line of the program, and the column on the right gives some explanation of the command.

Instruction	Explanation
PROGRAM:GAMBLER	Press [PRGM], select the **NEW** menu, and press [ENTER]. Name the program **GAMBLER** and press [ENTER].
: ClrHome	Clears the home screen. Press [PRGM], select the **I/O** menu, and scroll down to **8:ClrHome**.
: 0→S : 0→D	Variable S counts the number of triplets followed by the same outcome; variable D counts the triplets followed by a different outcome. Set both to a starting value of zero using the [STO▸] key.
: ClrList L5	Clears any data in list 5. Press [STAT] and select the **EDIT** menu for **ClrList**. Then press [2nd] [L1].
: Input "FLIPS?",F	Asks how many flips of the coin to perform; saves that value as F. Press [PRGM] and select the **I/O** menu for **Input**.
: randInt(1,2)→L5	These two lines randomly generate N flips, where N goes from 1 to T, and store the results in list 5. Press [PRGM] and select the **CTL** menu for **Pause** (scroll down to find **Pause**).
: Pause L5	
: For(K,4,F) : L5(K-1)→A : L5(K-2)→B : L5(K-3)→C : If A=B : Then : If A=C : Then	These seventeen lines start by looking at the fourth flip and checking whether the three flips before it form a triplet. If they do, the fourth flip is compared to the triplet and either variable S or D is increased by 1. If the previous three flips don't form a triplet, nothing happens. The group of commands then repeats by looking at the fifth flip, and so on. Press [PRGM] and select the **CTL** menu for **If**, **Then**, and **Else**. Press [2nd] [TEST] and select the **TEST** menu for =.

Continued on next page

```
: If L5(K)=A
: Then
: S+1→S
: Else
: D+1→D
: End
: End
: End
: End
: Disp "SAME:",S
: Disp "DIFFERENT:",D
```

The program displays the number of triplets followed by the same or a different outcome. Press [PRGM] and select the I/O menu for **Disp**.

When you run the program, the **Pause** command freezes the program so that you can see the data in list 5. Press the right and left arrow keys to inspect the data. When you are ready, press [ENTER] to resume the program.

Making a Frequency Bar Graph

You can build frequency bar graphs on your TI calculator. These bar graphs are convenient because as you change or correct your data set, the graph will automatically adjust to reflect the change. But beware of large data sets. A set of data from your entire class could exceed your calculator's storage capacity. The TI-83/84 family of calculators can hold up to 999 entries in each list.

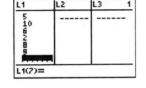

To find list **L1**, press [STAT] and [ENTER]. If list **L1** already contains data, put your cursor on the label of the list name and press [CLEAR] [ENTER]. Enter your data set.

To set up your calculator to build a frequency bar graph, press [2nd] [STAT PLOT]. You should get a STAT PLOT screen like the one shown here. You will use Plot 1. Press [ENTER] to display the Plot 1 screen.

Make your Plot 1 screen match the one shown here by highlighting **On**, and frequency bar graph, and pressing [ENTER] after each. Press [2nd] **L1** because this is where you stored your data set. In the last line highlight a frequency type 1.

Before you press [GRAPH], check a few things. First press [2nd] [STAT PLOT] and make sure all the other plots are turned off. Then, press [Y=] and make sure any functions are either deleted or turned off. You don't want any extra graphs cluttering your screen. Now press [GRAPH].

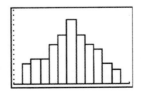

You will probably need to press [WINDOW] and adjust your window values to get a better view of your bar graph. If your window is the wrong size, you might get an error message when you try to graph. Select the window values carefully. **Xscl** is an important entry because it determines the width of each bar in your graph. In this example **Xscl** is set to 1. Make sure the **Ymax** value is big enough to accommodate the tallest bar in your graph.

Continued on next page

Reading the Frequency Bar Graph

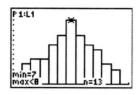

To read values from the frequency graph, press [TRACE]. Use the left and right arrow keys to move the cursor to different bars. Observe the numbers at the bottom of the screen. In the screen shown here, you can see the cursor on the tallest bar. Notice that the screen shows **min=7** and **max<8**. This means the cursor is on the bar containing *x*-values greater than or equal to 7 and less than 8. (Any value occurring on the edge of a bar is counted toward the bar on the right, so this bar contains all the 7s, but none of the 8s.) The calculator displays **n=13**, indicating that there are 13 entries in this bar.

Simulating Rollin' Rollin' Rollin'

You can combine what you've learned about random number generators and frequency bar graphs to simulate a large number of trials for *Rollin' Rollin' Rollin'*.

First press [STAT] and [ENTER], and clear the contents of lists L1, L2, and L3.

Arrow up to the name of list L1 and enter the random number generator **randInt(1,6,200)**. Do the same for list L2. These two lists now represent the outcomes on each of two dice for 200 rolls of the pair.

Arrow up to the name of list L3 and enter L1+L2. List L3 now represents the sum of the dice for 200 rolls.

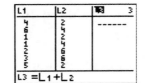

Now set up your frequency bar graph to use list L3. Compare your simulated frequency bar graph with a classmate's. How are they similar? How are they different?

You can experiment with a number of trials greater than 200, but beware of exceeding your calculator's storage capacity.

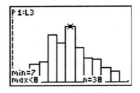

Pick a Card

A standard deck of cards contains four suits: clubs, diamonds, hearts, and spades. Each of these suits has 13 cards in it. The 13 cards include the 10 cards numbered 1 through 10, and 3 cards called jack, queen, and king. Imagine picking a card randomly from a deck of these cards.

You can simulate this event with a calculator program. The program steps given here are relatively simple and produce outputs like the examples in the display above. The two cards drawn in this example are the jack of clubs and the 2 of spades. (Can you see why the 11 indicates a jack?) If you want a challenge, stop reading now and write a program for this simulation without using the instructions below.

To follow these instructions, start a new program, give it a name of no more than seven characters, and enter the commands from the column on the left. The column on the right explains each command. You might try to improve the program by including display lines to indicate "ace" instead of 1, "king" instead of 13, and so on.

Instruction	Explanation
: randInt(1,13)→N	Chooses a random whole number from 1 through 13 and stores it in a cell labeled the variable N. Find **randInt** in the [MATH] **PRB** menu. Use the [STO▶] key for the → symbol.
: Disp N	Displays N. Find **Disp** in the [PRGM] **I/O** menu.
: randInt(1,4)→S	Chooses a random integer from 1 through 4 and stores it in a cell labeled the variable S.
: If S=1	Find **If** in the [PRGM] **CTL** menu. Find = by pressing [2nd] [TEST]. If the condition in an "IF" instruction is true, the calculator carries out the next instruction. If the condition is false, the calculator skips the next instruction.
: Disp "CLUBS"	Displays CLUBS if S is 1.

Continued on next page

The Game of Pig

`:If S=2`	
`:Disp "DIAMONDS"`	Displays DIAMONDS if S is 2.
`:If S=3`	
`:Disp "HEARTS"`	Displays HEARTS if S is 3.
`:If S=4`	
`:Disp "SPADES"`	Displays SPADES if S is 4.

Simulating One-and-One

The activity *One-and-One* describes the rules for a basketball free throw situation. The instructions here describe a calculator program that simulates this situation. A calculator running this program can produce many random outcomes of the game quickly. The display shown here gives two outcomes of the simulation.

To create the program, start a new program, give it a short and appropriate name such as **ONENONE**, and enter the instructions from the column on the left. The column on the right explains the function of each programming instruction.

Instruction	Explanation
: 0→T	Stores 0 in a cell labeled variable T. This will keep track of the total score of the free throws. Use [STO▶] for the → symbol.
: Lbl 1	Places a label at the start of the program. If the player scores on the first throw, the program will return to this label to begin the simulation of a second throw. Find **Lbl** in the [PRGM] **CTL** menu.
: randInt(1,10)→S	Chooses a random integer between 1 and 10 and stores this number in S.
: If S>6	Find > in the [2nd] [TEST] menu. Find **If** in the [PRGM] **CTL** menu.
: Goto 5	If the random number S was 7, 8, 9, or 10 (a 40 percent chance), the basketball player missed the throw and the simulation proceeds to label 5 near the end of the program. Find **Goto** in the [PRGM] **CTL** menu.
: T+1→T	If the random number S was 6 or less, the simulation adds one point to the total score stored T.
: If T=2	Find = in the [2nd] [TEST] menu.
: Goto 5	If the score is 2, the game stops, and the simulation proceeds to label 5 at the end of the program.

`:Goto 1`	If the score is not yet 2, the player gets another throw, and the simulation proceeds to label 1 at the beginning of the program.
`:Lbl 5`	Places a label at the end of the simulation.
`:Disp "TOTAL SCORE IS",T`	Displays these words before the final score.

As an added challenge, use what you've learned from other programs to make this simulation fully functional for *A Sixty Percent Solution*. Can you find a way to keep track of Terry's total scores for several trials? Can you use that list of scores to find Terry's most frequent outcome and average score in the long run? (The program ONEONE2.8xp is one example.)

Programming the Carrier Simulation

A Fair Deal for the Carrier? describes a simpler version of the newspaper carrier's dilemma from *The Carrier's Payment Plan Quandary*. These pages contain directions for programming a simulation of the carriers situation into your calculator. The program simulates 100 weeks of randomly drawing $20 or $1 bills.

The screen shown here displays a sample output of this program. The carrier has collected $518 over 100 trials. This means he collects an average of $5.18 per week. In this case picking a bill from the bag was a slightly better deal then simply taking the $5.

To simulate picking randomly from the bag, the program uses the **rand** command. Because there are five bills in the bag, the probability of picking the $20 bill is .2. Any random number less than .2 represents picking the $20 bill, and any random number of .2 or more represents picking one of the $1 bills.

To create the simulation described here, start a new program on your calculator and give it an appropriate name, like **CARRIER**. Enter the instructions from the column on the left into your calculator. The column on the right explains each instruction and gives hints about where to find some of the trickier commands.

Instruction	Explanation
:0→M	Starts M at zero. The variable M represents the total amount of money collected. The [STO▸] key gives you the → symbol.
:For(N,1,100)	Marks the place in the program where the experiment repeats for 100 trials. Find **For** in the [PRGM] **CTL** menu. Each time the experiment runs, N increases by one from 1 to 100.
:rand→R	Selects a random decimal and stores it as the variable R. Find **rand** in the [MATH] **PRB** menu
:If R<.2	Find < in the [2nd] [TEST] menu. Find **If** in the [PRGM] **I/O** menu.

`:M+20→M`	These last two instructions together mean that if you pick the $20 bill, increase your total money M by $20. If the condition in an **If** instruction is true, the calculator carries out the next instruction. Otherwise, the calculator skips it.
`:If R≥.2` `:M+1→M`	These last two instructions mean that if you pick the $1 bill, increase your total money M by $1.
`:End`	If N is less than 100, the simulation should return to **For** and run another trial. Computer programmers call this a loop because the program loops back to an earlier set of instructions. Find **End** in the [PRGM] **CTL** menu.
`:Disp "TOTAL COLLECTED",M`	Displays the total collected over all 100 trials. The simulation reaches these instructions only when N has reached 100.
`:Disp "AVG PER TRIAL", M/100`	Displays the average result of the 100-trial experiment.

Optional: Improving the Simulation

The simulation does not always have to run 100 trials. Try improving the simulation to allow the user to determine the number of trials. You also may have thought of some other improvements you would like to see.

A Simulation for The Game of Pig

Enter the program into a TI calculator to simulate the game of Big Pig. If you would prefer to simulate Little Pig, follow the directions in the right-hand column that explain the few modifications you need to make along the way.

The program allows you to select the type of strategy you use and the number of games you play. A display like the one shown here will prompt you to choose a strategy that limits either the number of rolls (or draws) per turn or the number of points per turn.

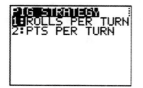

The calculator will also prompt you to input the number of turns it should play. When the calculator finishes all the games, it displays the total points from all the turns and the average score per turn.

Warning: The number of turns you will need to play in order to accurately determine the best strategy is quite high, perhaps as many as 200,000 turns! This is many more turns than is practical on any single TI calculator. However, you can see some general trends in the success of different strategies. (You can also play a large number of turns if you use several calculators simultaneously.)

The steps for the program are in the column on the left. The column on the right explains the steps. The text beneath each gray bar helps to explain different parts of the program.

Instruction

Program: PIG
: 0→B:0→T:0→Z

Explanation

Sets these variables equal to zero. The [STO▶] key gives you the → symbol. You can join several instructions into one line using a colon, [ALPHA] [:], above the decimal point key.

Continued on next page

`: ClrHome`	Clears the home screen. Find **ClrHome** in the PRGM I/O menu.
	The next line created a menu that enables you to select between two strategies. One strategy is to limit the total number of points; the other is to limit the number of rolls. Based on your choice, the program sends the calculator to the subroutine that matches the strategy at label 0 or label 1.

`: Menu("PIG STRATEGY",`
`"ROLLS PER TURN",0,`
`"PTS PER TURN",1)`
	If you are programming Little Pig, use **DRAWS** instead of **ROLLS**. Find **MENU** in the PRGM PRB menu.

	The next section begins the subroutine for the "limited-number-of-rolls" strategy. The calculator will ask first for the number of rolls you want to use for this strategy and then for the number of turns you want to play.
`: Lbl 0` `:Input "NO. OF ROLLS?",D`	Find **Lbl** in the PRGM CTL menu. If you are programming Little Pig, use **DRAWS** instead of **ROLLS**. Find **Input** in the PRGM I/O menu.
`: Input "NO. OF TURNS?",G`	Asks you to select a total number of games to play; stores that value in G.
`: Lbl 2`	
`: 0→A`	
`: Lbl 3`	These labels mark the beginning of the two loops in the "rolls" subroutine.
	The next section begins the games, using the variable Z to keep count of the number of turns played. This section also simulates the roll of the die, keeps track of the points for the turn, watches for a roll of 1, and keeps track of the total for all turns played.

Continued on next page

`: randInt(1,6)→P`	Selects a random number for the die; enters the value into P. Find **randInt** in the [MATH] **PRB** menu. If you are programming Little Pig, replace this line with `: randInt(1,3)→C`.

`: If P=1`	Find **If** in the [PRGM] **CTL** menu. Find = in the [2nd] [TEST] menu. If you are programming Little Pig, replace this line with these five lines:
	`: If C=2`
	`: 1→P`
	`: If C=3`
	`: 4→P`
	`: If C=1`

`: Goto 8`	If a 1 is rolled, sends the calculator to the subroutine to end the turn and give it a total of 0 points. Find **Goto** in the [PRGM] **CTL** menu.
`: T+P→T`	For any roll other than 1, adds the value to the total for that turn.
`: Lbl 4`	Gives a marker for the "rolls" subroutine to return to.
`: A+1→A`	Increases A by 1. The variable A counts the total number of rolls so far.
`: If A<D`	
`: Goto 3`	Checks to see if the number of rolls is still less than the number you chose. If so, the program returns to label 3 to roll again.
`: T+B→B`	When A exceeds D, accumulates the total points from all the turns in B.
`: 0→T`	Resets the points in T to 0 for the next turn.
`: Z+1→Z`	Increases Z by 1. The variable Z keeps track of the total number of games.
`: If Z≤G`	Tests to see if Z has exceeded G. The variable G represents the total number of turns you chose to play.
`: Goto 2`	If Z is still less than G, this sends the calculator to label 2 to begin another turn.
`: Goto 10`	Sends the calculator to the end-of-the-game subroutine to display the total points and the expected value.

Continued on next page

The Game of Pig

IMP Year 1: Calculator Notes for the TI-83/84 Family of Calculators

> This section begins the subroutine to play the game according to a "limited-number-of-points" strategy.

`: Lbl 1`	Begins the subroutine if you selected a "points" strategy.
`: Disp "NO. OF POINTS?",D`	Asks you to pick a maximum number of points for each turn; stores this value in D.
`: Input "NO. OF TURNS?",G`	Asks you to select a total number of turns; stores that value in G.
`: Lbl 5`	Marks the beginning of all the turns for the "points" strategy.
`: Lbl 6`	Marks the beginning of a single turn using the "points" strategy.
`: randInt(1,6)→P`	Selects a random number for the die; enters the value into P. If you are programming Little Pig, replace this line with `: randInt(1,3)→C`.

`: If P=1`	If you are programming Little Pig, replace this line with these five lines:
	`: If C=2`
	`: 1→P`
	`: If C=3`
	`: 4→P`
	`: If C=1`

`: Goto 9`	If a 1 is rolled, sends the calculator to the subroutine to end the turn with 0 points.
`: T+P→T`	Adds the value of each roll to the total for the turn.
`: T→A`	Assigns the value of T to the variable A to see if the maximum for this turn has been reached.
`: Lbl 7`	Gives a marker for the calculator to return to if a 1 is rolled.
`: If A<D`	Checks to see if the total points is still less than the points limit in your strategy.
`: Goto 6`	Sends the calculator to label 6 to roll again, assuming the total point was still less than D.
`: T+B→B`	Keeps track of the accumulated total for each turn.

Continued on next page

Interactive Mathematics Program © 2009 Key Curriculum Press

`: 0→T`	
`: 0→A`	Resets the variables to 0 to begin the next turn.
`: Z+1→Z`	Adds 1 to Z. The variable Z keeps track of the number of turns played so far.
`: If Z≤G`	Tests to see if Z has exceeded G. The variable G represents the number of turns you chose to play.
`: Goto 5`	Sends the calculator to label 5 to begin another game if Z is still less than or equal to G.
`: Goto 10`	Sends the calculator to the end-of-the-game subroutine to display the total points and the expected value.

This section resets the points if a 1 is rolled while using a "limited-number-of-rolls" strategy.

`: Lbl 8`	Begins the subroutine to erase points and end a turn if a 1 is rolled.
`: D→A`	Sets the number of rolls to the maximum in order to end the turn.
`: 0→T`	Sets the points for the turn to 0.
`: Goto 4`	Returns the calculator to the "rolls" subroutine.

This section resets the points if a 1 is rolled while using a "limited-number-of-points" strategy.

`: Lbl 9`	Begins the subroutine to erase points and end a turn if a 1 is rolled.
`: D→A`	Sets the number of points to the maximum in order to end the turn.
`: 0→T`	Sets the points for the turn to 0.
`: Goto 7`	Returns the calculator to the "points" subroutine.

Continued on next page

The Game of Pig

IMP Year 1: Calculator Notes for the TI-83/84 Family of Calculators

---

> This section tells the calculator to display the total number of points for all the games played and the average score for this particular strategy and displays the results.

`:Lbl 10`	Begins the subroutine to display total points and average score.
`:Disp "TOTAL POINTS:",B`	Displays the total points from all the turns. Find **Disp** in the [PRGM] I/O menu.
`:Disp "AVERAGE SCORE:",B/G`	Displays the average score per turn.
`:Stop`	Ends the entire program. Find **Stop** in the [PRGM] CTL menu.

This chart contains expected values for some specific combinations of turns and strategy. You can use this information to help check whether your simulation is running correctly. You should get averages within the "reasonable" range described in the table practically every time.

	Little Pig	Big Pig
Limiting number of draws or rolls	3 draws, 500 turns Expected value = 2.222… Any value between 1.9 and 2.5 is a reasonable average.	3 rolls, 500 turns Expected value = 6.9444… Any value between 6.65 and 7.35 is a reasonable average.
Limiting number of points	3 points, 500 turns Expected value = 2.222… Any value between 2.0 and 2.4 is a reasonable average.	3 points, 500 turns Expected value = 3.8333… Any value between 3.6 and 4.0 is a reasonable average.